# Table of Contents

Y0-EKS-156

\* Full-color transparencies are found at the back of the book. Each transparency should be used to introduce the corresponding unit.

## Teaching Guide

This book has been designed to strengthen map reading and analyzing skills and to familiarize students with the features of different regions of Asia. Each section contains a student page and an accompanying map that is used to complete the student page. Each page builds upon previous pages and students should have access to all previously completed maps. Encourage students to use latitude and longitude markings when comparing features between maps. Students may use related resources to assist them in completing the student pages. Accept all reasonable answers that can be logically supported. Color transparencies, located in the back of the book, will assist you in your classroom presentation.

## Map 1:  Asia in the World

This Robinson projection map provides a continuous map with relatively little distortion of shapes and sizes. It is a compromise between rectangular projections such as the Mercator or Peters and more accurate interrupted projections. The student page offers directional activities and a world overview.

### Extension Activities:

1.  Note that Asia is the world's largest continent and compare its size to other more familiar areas. Compare the latitudes that Asia covers and discuss the effect of latitude on climate.

2.  Use this map to discuss the European exploration of the world. Have students trace the routes of Dias, Da Gama, and Magellan.

3.  Discuss non-European exploration and trade routes. Have students trace routes of Arabic, Indian, and Chinese traders.

## Map 2:  Countries and Capitals

This map is based on information accurate as of late 2008. Names change occasionally as political regimes rise and fall. Use this information with caution. This page can be used for latitude and longitude practice or the map can be labeled using a classroom atlas or textbook.

### Extension Activities:

1.  Make a copy of the map. Have students color the countries according to dates when independence was achieved. Use different colors for specified date ranges.

2.  Research the ownership of Asian colonies as of 1914. Color the colonies according to their controlling country: Netherlands, England, United States, France, and so on.

3.  Make choropleth maps for a variety of statistics. A choropleth map uses a different color for a different range. For example, have students research the literacy rate for the countries of the continent and then color the countries according to specified percentages (20%, 20-50%).

## Map 3:  Physical Features

The student page provides a narrative description of the features of Asia. Students will read the description and label the features on the map. Alternatively, the teacher may wish to number the features on the map and have students match the numbers with the feature names. The map has not been numbered to provide maximum flexibility.

### Extension Activities:

1.  Have students draw the borders of the continental plates. This information can be found in an encyclopedia or atlas.

2.  Have students use an atlas to find the latitudes and

longitudes of several high peaks (Everest, K-2, Annapurna, Fujiyama, and so on) and to locate them on their maps.

3. Research early trade and exploration routes and trace them on the map.

4. Have students label two copies of the map, one with water features and one with physical features.

## Map 4: Elevations
This map may be used as a companion to the physical features map. Students should understand that Asia's elevation is extremely diverse compared to continents such as Africa. The student page provides practice in reading an elevation map and comparing it to additional maps. It also provides computational and graphing practice.
### Extension Activity:
Students may make three-dimensional maps with clay, papier-mâché, etc. Alternatively, they can layer cardboard cut to correct dimensions to represent different elevations. For example, use one layer for every 100 meters or 500 feet of elevation.

## Map 5: Precipitation
This map provides an opportunity to discuss the effects of landforms and ocean currents on continental precipitation. For example, the Plateau of Tibet is dry for several reasons, such as rising air, distance from large bodies of water, and a range of very high mountains between it and the coast. The student page provides practice in locating geographic areas and countries and comparing maps.
### Extension Activity:
Use an almanac to find exact precipitation records for Asian cities. Construct bar or line graphs to compare cities in different areas.

## Map 6: Climates
Review the different climate types with the students. It is important to remember that geographers use different classification systems. This is a simplified system.
1. **Tropical rainy**–areas with hot, wet climate all year
2. **Tropical wet-dry**–hot regions that receive most of their rain during one period of the year followed by a period of dry weather
3. **Semi-arid or steppe**–areas that receive relatively little rainfall and are subject to drought and possible desertification
4. **Desert**–areas that receive little if any rainfall and have hot days and cool to cold nights
5. **Mediterranean**–regions that have constantly mild temperatures that do not vary greatly from summer to winter and have moderate precipitation
6. **Highland**–regions that vary greatly in rainfall but always have lower temperatures than the surrounding areas
7. **Humid subtropical**–wet areas with cool winters and warm summers
8. **Humid continental**–moderately wet areas with cold winters and cool to warm summers
9. **Tundra**–areas that have cool to cold summers and extremely cold winters

## Map 7: Natural Vegetation
This map does not pertain to agricultural regions but to naturally occurring vegetation types. It correlates with the climate and precipitation maps, enabling students to gain an appreciation of these relationships. The student page provides locational and comparative practice. The teacher should decide whether metric or English measurements will be used.
### Extension Activities:
1. Have students research the types of plants that comprise each of these vegetative zones. Discussions or reports should center around biogeographical questions such as, "Why does this type of plant grow only here?"

2. Students can research the animal life of Asia and relate this to the vegetative and climate zones. Again, biogeographical questions can be asked: Why are these animals found here? What vegetation do they need? How might their environment be threatened?

## Map 8: Products
Asia is a continent of great resources, many of which have been used wisely for many centuries. The map shows both cultivated products and mineral resources.
### Extension Activities:
1. Students should familiarize themselves with the products described on the map.

2. Research Asia's importance as a world supplier of raw materials and minerals. Students can trace the routes of products as they are changed from raw materials to finished goods.

## Map 9: Land Use
This generalized map indicates the predominant land use activity over a very large area and is useful for comparison activities. Students should be careful when drawing conclusions about land use. For example, there are few similarities between the type of farming practiced on the steppes of Russia and the intensive agriculture of east and southeast Asia. It is interesting to compare this map with the climate, natural vegetation, and products maps and to draw conclusions about relationships.
### Extension Activities:
1. Have students use almanacs or encyclopedias to determine the percentages of arable land in each country. Create a choropleth map showing the results. (See map 2.)

2. Discuss how changing land use patterns in individual countries, such as Indonesia, are threatening the world's rainforests.

## Map 10: Population Patterns and Large Cities
Density refers to the average number of people per square kilometer or mile. Parts of Asia are populated to the point (and beyond) of overpopulation. Areas that are not as densely populated are probably low in arable land or high in elevation. Teachers should discuss the reasons for population variance keeping in mind climate, precipitation, physical features, and political borders.
### Extension Activities:
1. Research different countries and write a paragraph

explaining the reasons for their population patterns.

   2. Consult an almanac to determine the present population of large cities. Label any cities that have grown to more than two million people on the map.

   3. Consult an almanac. Compare population densities with birthrates and life expectancies.

## Map 11: Empires and Civilizations

This map depicts the areal extent of selected kingdoms, cultures, and empires and is keyed to show approximate time frames for their climax periods. This information is given to build a sense of appreciation for how empires emerge in the same geographic areas. The accompanying student page includes map reading and analyzing questions.

### Extension Activities:

   1. Research the cultures and create time lines based on these cultures.

   2. Research trading routes that connected these cultures with each other and with outside cultures.

   3. Plan imaginary trading routes between two or more contemporary cultures. Use additional maps to describe the areas traveled.

   4. Investigate how warfare and empire building has changed political boundaries and influenced cultures. Note how certain civilizations keep re-emerging despite periods of subjugation (e.g., Persia, China).

## Maps 12-16: Regional Maps

These maps provide opportunities for detailed map study or for regional studies of Asia. The student pages provide labeling activities, latitude and longitude practice, direction and scale practice, and practice in comparative geography. The maps may be used in a variety of teacher-directed activities.

### Extension Activities:

   1. Color the maps to make the country borders more obvious and to make the map more aesthetically pleasing.

   2. Label the maps with climate, elevation, or precipitation information.

   3. Use these maps to plot past military campaigns.

## Answer Key

**Page 1** 1-5 Consult an atlas or Milliken transparency for answers. 6a. Pacific b. north c. south d. northern, eastern e. Africa f. west g. Atlantic h. south i. Indian j. Australia

### Page 2

Southern Asia

| | |
|---|---|
| 12 | 27 |
| 25 | 47 |
| 29 | 40 |
| 3 | 30 |

Eastern Asia

| | |
|---|---|
| 2 | 28 |
| 11 | 45 |
| 15 | 16 |

Southeastern Asia

| | |
|---|---|
| 32 | 18 |
| 44 | 8 |
| 34 | 39 |
| 36 | 41 |
| 24 | 4 |

Northern Asia

| | |
|---|---|
| 5 | 43 |
| 23 | 7 |
| 21 | 13 |

Southwestern Asia

| | |
|---|---|
| 20 | 35 |
| 31 | 42 |
| 6 | 46 |
| 33 | 37 |
| 22 | 14 |
| 1 | 48 |
| 26 | 19 |
| 17 | 38 |
| 9 | 10 |

**Page 3** Consult an atlas or Milliken transparency for answers.

**Page 4** 1. 700-3000 ft. 2. sea level to 700 ft. 3. 700-3000 ft. 4. 700-3000 ft. 5. sea level to 700 ft. 6. 3000-7000 ft. 7. over 13,000 ft. 8. below sea level 9. sea level to 700 ft. 10. 7000-13000 ft. 11. sea level to 700 ft. 12. 700-3000 ft. 13. sea level to 700 ft. 14. Himalayas 15. Caspian, Aral, sea level to 700 ft. 16. Tigris and Euphrates rivers, sea level to 700 ft. 17. Ganges River, sea level to 700 ft. 18. Iran 19. Indonesia 20. Nepal 21. Turkey 22. Afghanistan 23. Mongolia 24. Pakistan

**Page 5** 1. over 80 in. 2. 60-80 in. 3. under 10 in. 4. under 10 in. 5. 40-60 in. 6. 40-60 in. 7. under 10 in. 8. 60-80 in. 9. 10-20 in. 10. 20-40 in. 11. south or southeast 12. north, East 13. warm 14. Equator 15. warm, warm air holds more moisture 16. There are many warm ocean currents due to its location along the equator. 17. The air rising over the Himalayas causes precipitation to occur. 18. It is located on the back side of the Himalayas and the air has lost its moisture. 19. ocean currents moving away from land 20. under 10 in. 21. 10-20 in. 22. under 10 in. 23. 20-40 in. 24. over 80 in.

**Page 6** 1. tropical rainy, 20-40 in. 2. tundra, under 10 in. 3. tropical semi-arid/desert, under 10 in. 4. Tundra is much colder than desert or tropical dry. 5. humid subtropical, continental 6. tropical semi-arid/desert 7. mediterranean, 40° N 8. Cancer, Capricorn 9. subarctic 10. tropical rainy and humid subtropical 11. tropical rainy 12. continental 13. tropical wet-dry 14. tropical wet-dry or humid subtropical 15. tropical semi-arid 16. tropical wet-dry 17. e 18. a 19. b 20. f 21. c 22. d

**Page 7** 1. dry savannah 2. dry savannah 3. rainforest, mixed deciduous, desert 4. over 80 in. 5. tropical rainy 6. desert, highland 7. rainforest, over 80 in., because of the very high mountains 8. mediterranean 9. rain forest, mixed deciduous 10. southeast, Bay, Bengal, mixed deciduous 11. taiga 12. mediterranean, dry savannah, mediterranean 13. mediterranean, dry savannah, desert 14. very wet and warm climate due to presence of ocean 15. north, coniferous forest, tundra 16. d 17. c 18. e 19. b 20. a

**Page 8**

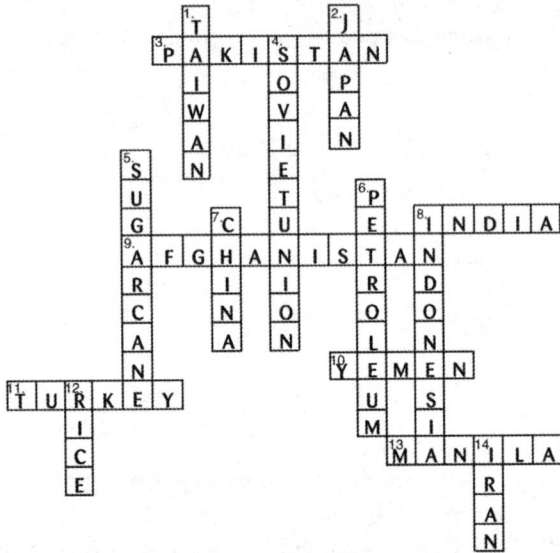

**Page 9** 1. nomadic herding, desert 2. steppe, nomadic herding 3. Huang He, Yangtze, rice, wheat, barley, corn 4. heavy rain, clearing forests 5. forest, farming, rice, sugarcane 6. nomadic herding, stock grazing 7. little used land, very hot and dry 8. high elevations, cold temperatures, little precipitation 9. goats, sheep, cattle 10. rubber, corn 11. farming, forest, fishing 12. steepness of hillsides, traditional patterns of land use, lack of rainfall, climate 13. need to support the heavy population by growing food 14. Nomads move from place to place, grazing is sedentary and remains in one place. 15. Farming produces the most food.

**Page 10** Consult an atlas or Milliken transparency for answers. 1. (choose 2) India, China, Japan 2. less than 1 per km, barren land, little rainfall, arid, desert climate 3. little vegetation, little rainfall, arid desert climate 4. River valleys tend to have more vegetation and adequate water to support larger populations. 5. the island of Luzon

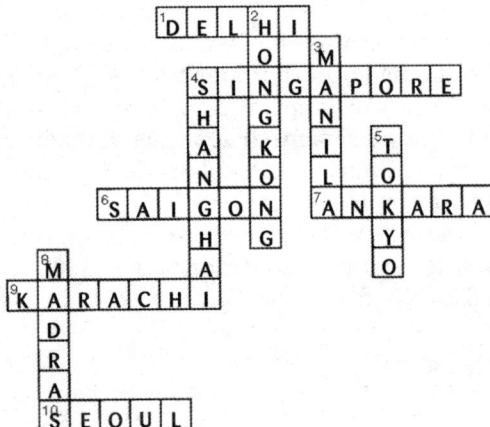

**Page 11** Consult additional resource or Milliken transparency for answers. 1. Chin, Han, Ming 2. (choose any three) Iran, Afghanistan, Iraq, Syria, Turkey, Lebanon, Israel, Kuwait 3. Roman, Muslim, Persian 4. It is isolated, as it is surrounded by water. 5. f 6. a 7. d 8. b 9. c 10. e

**Page 12** 1-5 Consult an atlas or Milliken transparency for anwers. 6. answers will vary 7. Japan, East China, Okhotsk 8. 1980km/1280 mi., southwest 9. north, Arctic Ocean 10. north, south

**Page 13** 1-3 Consult an atlas or Milliken transparency for answers. 4. Turkey, Israel, Israel, Iran, Iran, Saudi Arabia, Saudi Arabia, Syria 5. answers will vary 6. Marmara, Dardanelles, Aegean, Mediterranean, Suez, Red, Aden, Arabian, Oman, Hormuz, Persian 7. 150 km/110 mi., south 8. Rub al Khali

**Page 14** 1-3 Consult an atlas or Milliken transparency for answers. 4. Indonesia, Indonesia, Indonesia, Indonesia, Myanmar, Vietnam, Vietnam 5. answers will vary 6. south, Andaman, Malacca, Singapore, Java 7. 1132 km/704 mi., north 8. Thailand, Manila, east, 2190 km/1370 mi., South China

**Page 15** 1-3 Consult an atlas or Milliken transparency for answers. 4. South Korea, China, China, China, China, China, China, Japan, Japan 5. answers will vary 6. northwest, 1150 km/750 mi., Gobi 7. south, Yellow, East China 8. Plateau, Tibet, east 2900 km/1800 mi.

**Page 16** 1-3 Consult an atlas or Milliken transparency for answers. 4. Pakistan, Pakistan, Bangladesh, Nepal, Sri Lanka, India, India, India, India, India 5. answers will vary 6. 970 km/600 mi., southwest 7. Kathmandu, east, 800 km/500 mi., India 8. Arabian, Indian, Bengal 9. southeast, 1000 km/630 mi., Pakistan 10. India, Bangladesh, Bengal

# Asia in the World

Use the map on page 1a to label the features and to complete the following.

1.  Label and color the continents.
    | Africa | green |
    | Asia | orange |
    | Europe | yellow |
    | Australia | red |
    | North America | brown |
    | South America | purple |
    | Antarctica | gray |

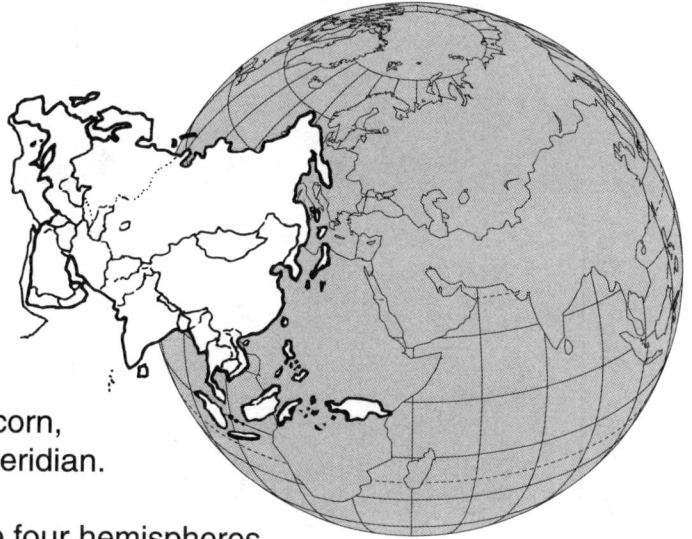

2.  Label these parallels and meridians:
    Equator, Tropic of Cancer, Tropic of Capricorn,
    Arctic Circle, Antarctic Circle, and Prime Meridian.

3.  Label the North and South poles. Label the four hemispheres.

4.  Label the oceans: Pacific, Atlantic, Indian, and Arctic.

5.  Label the compass rose on the map with these directions: north, south, east, west,
    northeast, southeast, northwest, and southwest.

6.  Use the eight major directions or names of geographic features to complete the following.

    a.  The _____ Ocean lies to the east of Asia.

    b.  The Arctic Ocean lies _____ of Asia.

    c.  South America lies east and slightly _____ of Asia.

    d.  Asia is located in the _____ and _____ hemispheres.

    e.  Asia is northeast of the continent of _____ .

    f.  Europe is _____ of Asia.

    g.  None of Asia borders the _____ Ocean.

    h.  Antarctica is _____ of Asia.

    i.  The _____ Ocean is south of Asia.

    j.  The continent of _____ is south and southeast of Asia and borders the
        Indian Ocean.

# Asia in the World

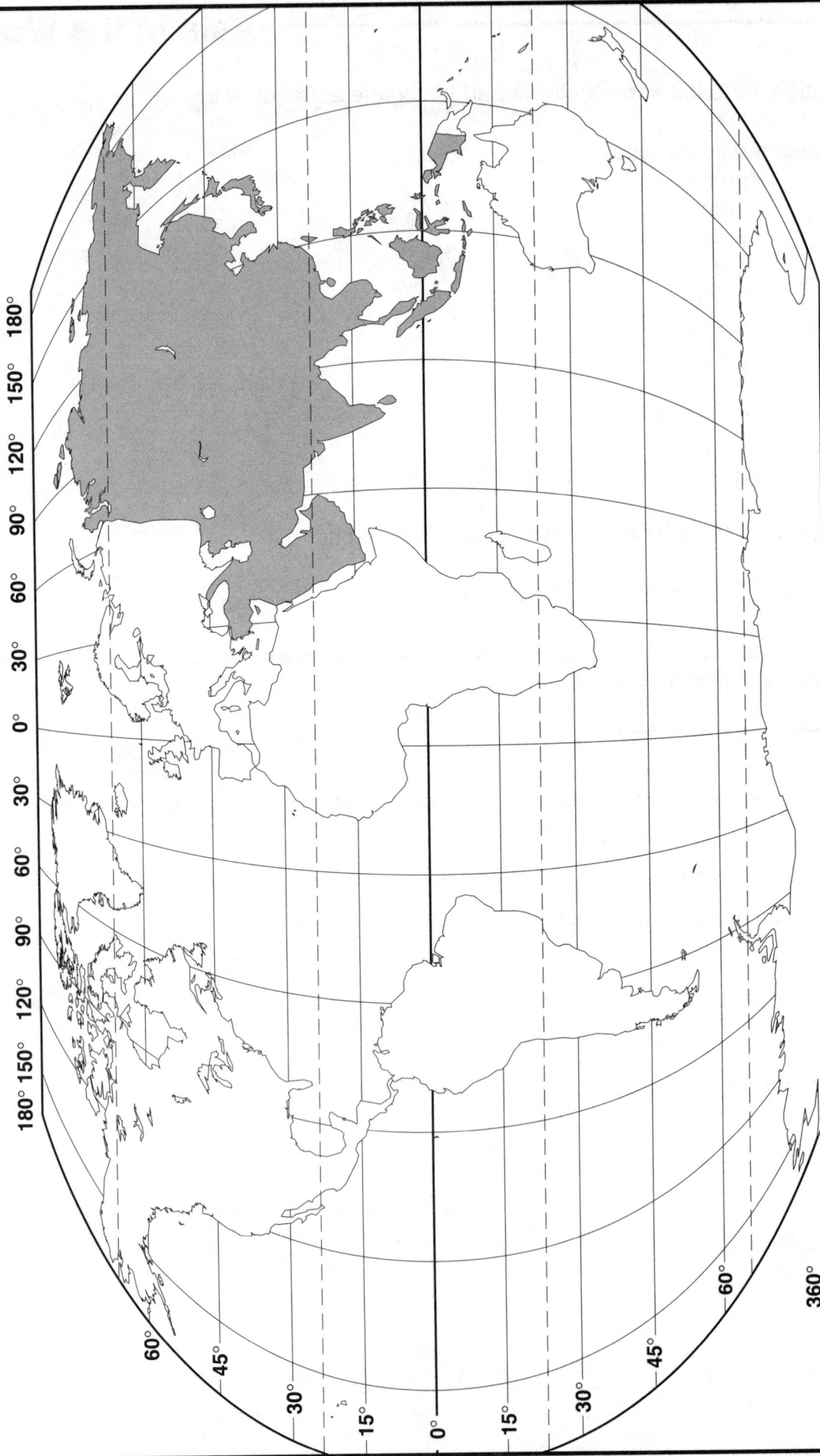

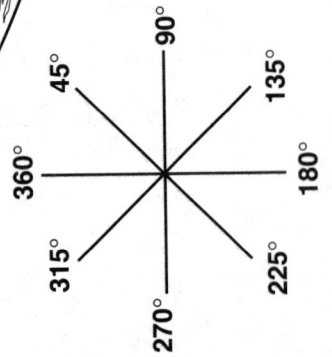

180° 150° 120° 90° 60° 30° 0° 30° 60° 90° 120° 150° 180°

60°
45°
30°
15°
0°
15°
30°
45°
60°

45°
90°
135°
180°
225°
270°
315°
360°

# Countries and Capitals

Match each country and its capital with the corresponding number on the map on page 2a.

| Capital, Country | Capital, Country |
|---|---|

## Southern Asia

| | | | |
|---|---|---|---|
| _____ | New Delhi, India | _____ | Kathmandu, Nepal |
| _____ | Islamabad, Pakistan | _____ | Thimphu, Bhutan |
| _____ | Dhaka, Bangladesh | _____ | Kabul, Afghanistan |
| _____ | Colombo, Sri Lanka | _____ | Male, Maldives |

## Eastern Asia

| | | | |
|---|---|---|---|
| _____ | Beijing, China | _____ | Tokyo, Japan |
| _____ | Ulaanbaatar, Mongolia | _____ | Pyongyang, North Korea |
| _____ | Taipei, Taiwan | _____ | Seoul, South Korea |

## Southeastern Asia

| | | | |
|---|---|---|---|
| _____ | Nay Pyi Taw, Myanmar (Burma) | _____ | Kuala Lumpur, Malaysia |
| _____ | Bangkok, Thailand | _____ | Singapore, Singapore |
| _____ | Vientiane, Laos | _____ | Jakarta, Indonesia |
| _____ | Phnom Penh, Cambodia | _____ | Manila, Philippines |
| _____ | Bandar Seri Begawan, Brunei | _____ | Hanoi, Vietnam |

## Northern Asia

| | | | |
|---|---|---|---|
| _____ | Moscow, Russia | _____ | Astana, Kazakhstan |
| _____ | Dushanbe, Tajikistan | _____ | Bishkek, Kyrgyzstan |
| _____ | Tashkent, Uzbekistan | _____ | Ashgabat, Turkmenistan |

## Southwestern Asia

| | | | |
|---|---|---|---|
| _____ | Ankara, Turkey | _____ | Baghdad, Iraq |
| _____ | Nicosia, Cyprus | _____ | Riyadh, Saudi Arabia |
| _____ | Damascus, Syria | _____ | Sanaa, Yemen |
| _____ | Beirut, Lebanon | _____ | Muscat, Oman |
| _____ | Amman, Jordan | _____ | Kuwait, Kuwait |
| _____ | Jerusalem, Israel | _____ | Doha, Qatar |
| _____ | Tehran, Iran | _____ | Manama, Bahrain |
| _____ | Abu Dhabi, United Arab Emirates | _____ | T'bilisi, Georgia |
| _____ | Yerevan, Armenia | _____ | Baku, Azerbaijan |

# Countries and Capitals

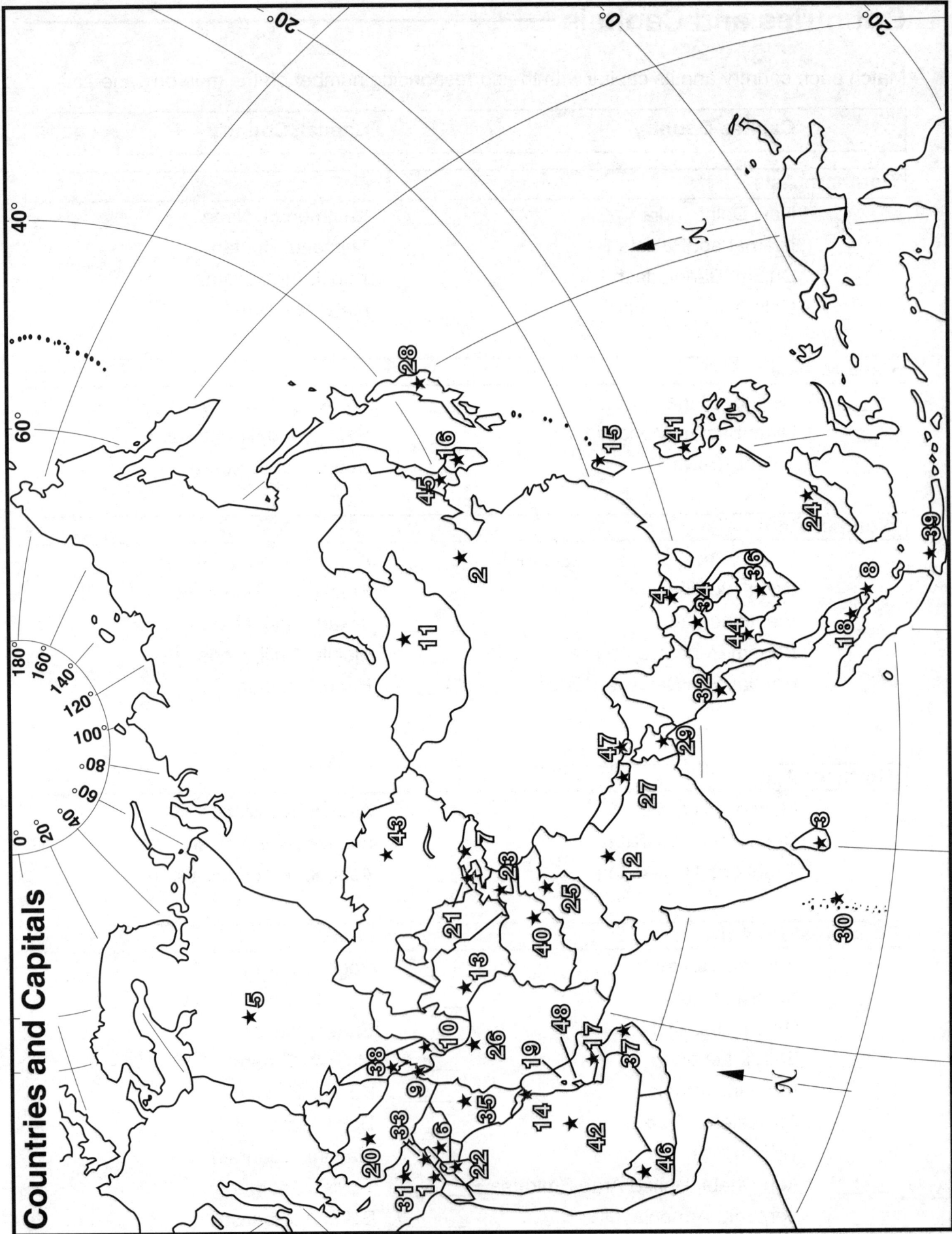

Read the following paragraphs. Label the following **boldfaced** physical features on the map on page 3a. You may consult previous maps or other sources for more information.

Many seas, bays, and gulfs surround the continent of Asia: **Sea of Japan,** 40°N, 135°E; **East China Sea,** 30°N, 125°E; **South China Sea,** 10°N, 110°E; **Yellow Sea,** 38°N, 120°E; **Philippine Sea,** 15°N, 130°E; **Bay of Bengal,** 15°N, 90°E; **Arabian Sea,** 20°N, 65°E; **Mediterranean Sea,** 33°N, 33°E; and the **Persian Gulf,** 27°N, 50°E.

A lake is a body of water surrounded by land. The world's largest lake is the **Caspian Sea,** 40°N, 52°E. The **Aral Sea** is located at 43°N, 60°E. **Lake Baikal,** the world's deepest lake, is located at 53°N, 108°E.

Many of the world's largest islands are considered part of Asia: **Honshu,** 36°N, 135°E; **Borneo,** 0°N, 112°E; the western portion of **New Guinea,** 5°S, 140°E; **Sumatra,** 0°S, 101°E; **Sakhalin,** 50°N, 142°E; **Java,** 7°S, 107°E; and **Luzon,** 15°N, 121°E.

Much of Asia is covered by high plateaus and mountains. The **Himalayas,** the highest mountain range in the world, is located between China and India. Just north of the Himalayas, in China, is the **Plateau of Tibet.** To the northwest of the Himalayas, at about 36°N, 70°W is the **Hindu Kush.** Northeast of the Hindu Kush along the Kyrgyzstan–China border is the **Tien Shan.** Forming part of the border between Europe and Asia are the **Ural Mountains,** which run north and south at 60°E longitude. Another part of the Europe–Asia boundary is formed by the **Caucasus Mountains,** which lie just west of the Caspian Sea. The **Zagros Mountains** are in western Iran and run northwest to southeast. East of these mountains is the **Iranian Plateau.** The **Deccan Plateau** is located in southern India. The highland area of Turkey is called **Anatolia.** The area just north of Lake Baikal is called the **Central Siberian Plateau.**

Plains, steppes, and deserts are also quite common in Asia. Just east of the Urals is the **West Siberian Plain.** The area between the Ural Mountains and the Aral Sea is known as the **Kirghiz Steppe.** Most of the land in southern Saudi Arabia is the desert known as **Rub' al Khali** (the Empty Quarter). Just southeast of the Tien Shan is a desert know as **Taklamakan.** Covering much of southern Mongolia is the large desert known as the **Gobi.** The **Great Indian Desert** is located along the border of India and Pakistan.

Many large rivers drain the continent. Siberia is drained by these rivers (going east from the Urals): the **Ob,** the **Yenisey,** and the **Lena.** The **Amur River** flows east along the Russia–China border. Two very large rivers flow east through China: the northern one, which flows into the Yellow Sea, is called the **Huang He** (Yellow River) and the southern one is the **Yangtze River.** Flowing from the Plateau of Tibet southeast to the South China Sea is the **Mekong River.** Flowing out of the Himalayas and into the Bay of Bengal is the **Ganges River.** Also flowing out of the Himalayas but into the Arabian Sea is the **Indus River.** The **Tigris** and **Euphrates Rivers** lie just west of the Zagros Mountains and flow from Anatolia southeast to the Persian Gulf. The Tigris is east of the Euphrates, which flows into the Tigris.

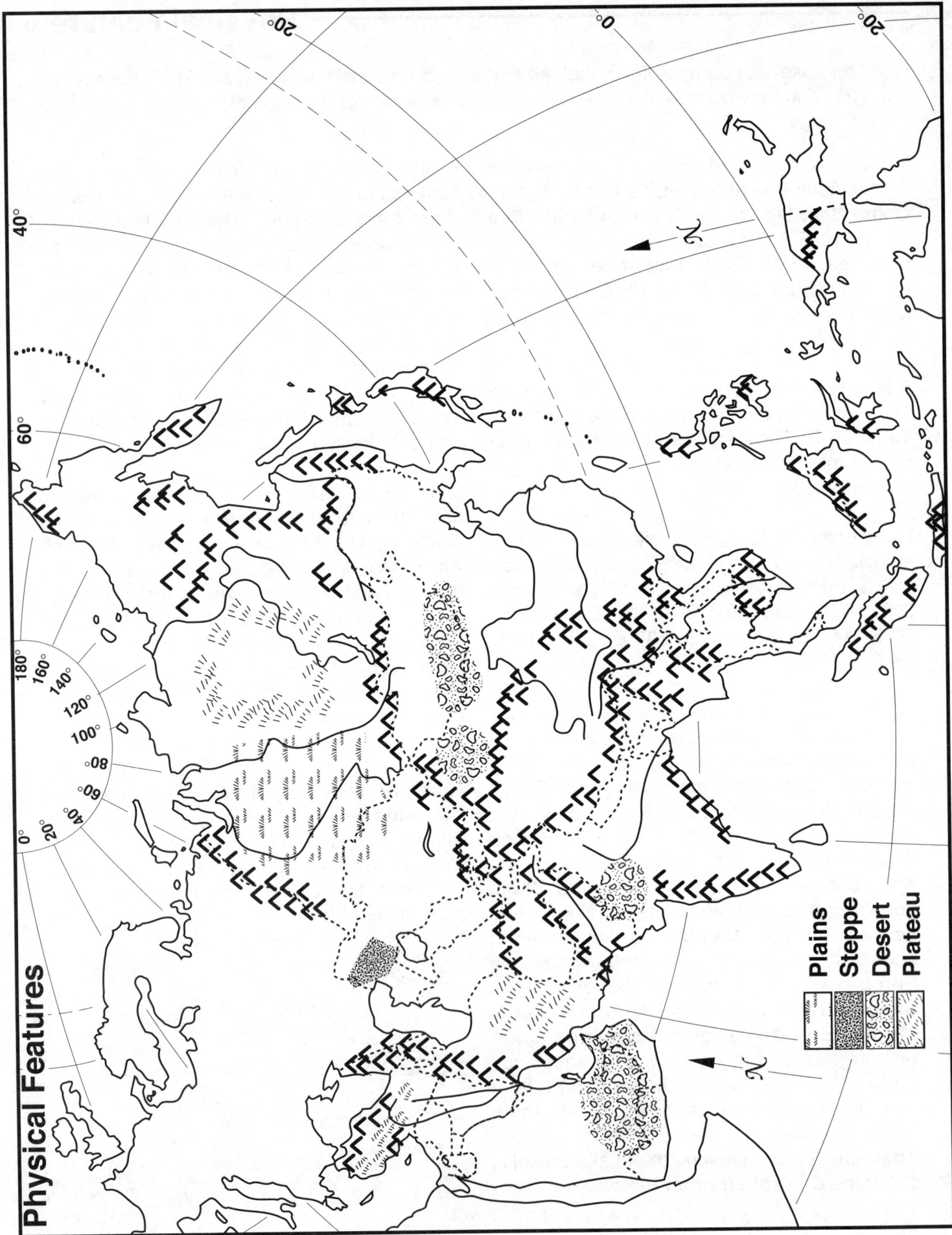

# Physical Features

**Plains**
**Steppe**
**Desert**
**Plateau**

# Elevations and Physical Features

Use the maps to complete the following.

List the elevation range for:
1. central Saudi Arabia _____
2. the Western Siberian Plain _____
3. the Central Siberian Plateau _____
4. the Deccan Plateau of southern India _____
5. Cambodia _____
6. most of Mongolia _____
7. most of the Plateau of Tibet _____
8. the east coast of the Caspian Sea _____
9. Baghdad, Iraq _____
10. Kabul, Afghanistan _____
11. Singapore _____
12. Amman, Jordan _____
13. Bangladesh _____

14. Name the landform responsible for the high elevations just north of India. _____

15. The Kirghiz Steppe lies northeast of the _____ Sea and northwest of the _____ Sea. What is its elevation range? _____

16. The area in eastern Iraq has long been referred to as Mesopotamia, which means "land between the rivers." To what two rivers does the name refer? _____
_____
What is the elevation range for most of Mesopotamia? _____

17. The lowland area in northeastern India is mainly drained by what river? _____
What is the elevation range for this area? _____

The term relief refers to the difference between the highest and lowest elevations of an area. List the country that has the greatest relief.
18. Iran or Saudi Arabia _____
19. the Philippines or Indonesia _____
20. Sri Lanka or Nepal _____
21. Turkey or Syria _____
22. Jordan or Afghanistan _____
23. Mongolia or Kuwait _____
24. Israel or Pakistan _____

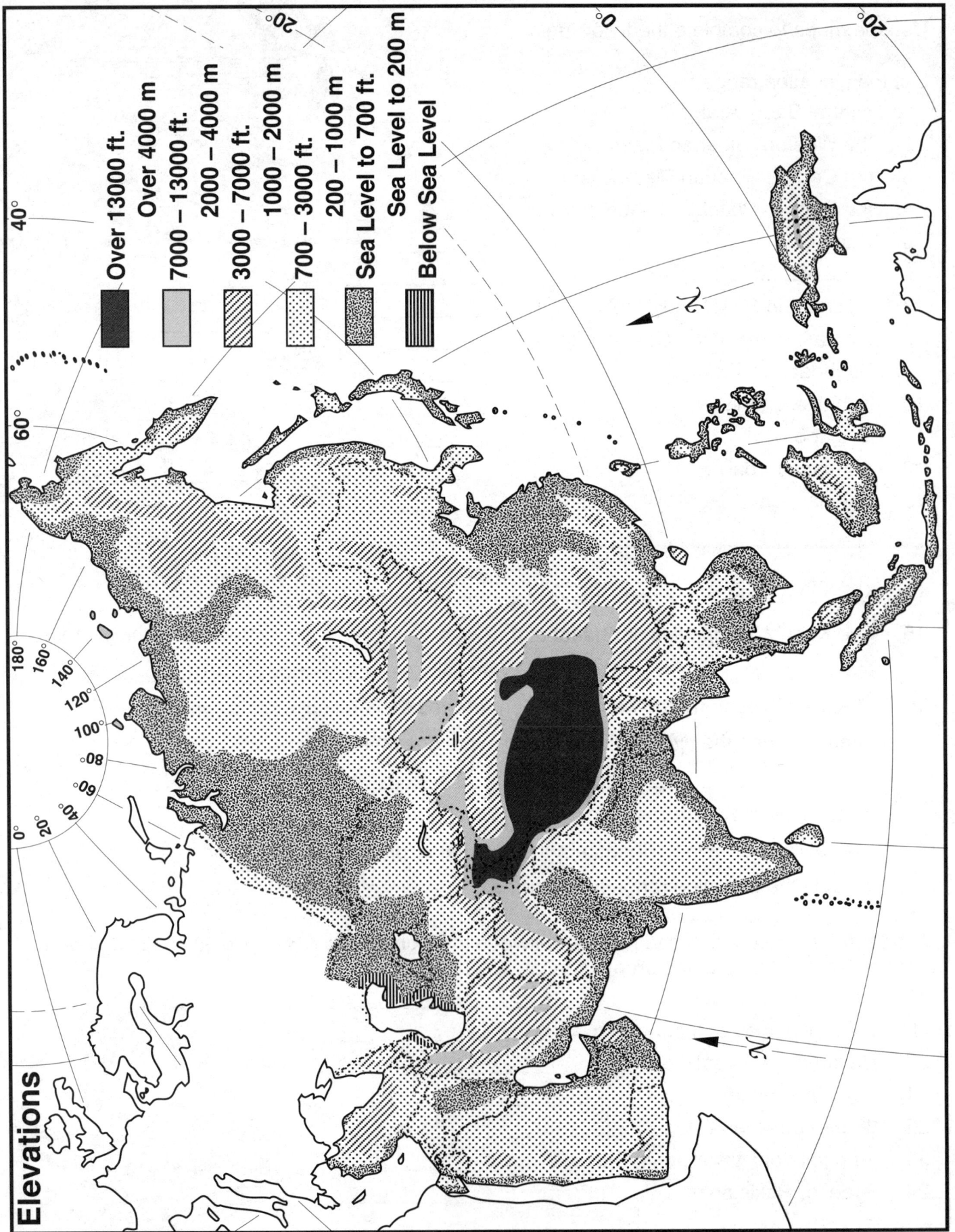

# Elevations

**Over 13000 ft.** / **Over 4000 m**

**7000 – 13000 ft.** / **2000 – 4000 m**

**3000 – 7000 ft.** / **1000 – 2000 m**

**700 – 3000 ft.** / **200 – 1000 m**

**Sea Level to 700 ft.** / **Sea Level to 200 m**

**Below Sea Level**

N

S

Use the maps to complete the following.

List the annual precipitation for:
1. most of Indonesia _____
2. southern Sri Lanka _____
3. central Saudi Arabia _____
4. central Mongolia _____
5. eastern South Korea _____
6. central Thailand _____
7. Kuwait _____
8. northern Sri Lanka _____
9. central Turkey _____
10. south-central India _____

11. Warm air currents move _____ along the west coast of India.
12. Currents between China and Taiwan flow _____ into the _____ China Sea.
13. Currents in the Sea of Okhotsk are mostly _____ .
14. Warm, circular currents are found along what special parallel? _____
15. Coastal areas of heavy precipitation are generally associated with _____ air currents. Why? _____
16. How does this help to explain the very heavy precipitation over most of the islands in southeastern Asia? _____

_____

17. When warm air rises over mountains, precipitation usually occurs. How does this explain the precipitation pattern in Nepal? _____
18. Why does the Plateau of Tibet receive so little precipitation? _____

_____

19. Why do the northwestern coastlands of the Arabian Sea receive so little precipitation?

_____

List the annual precipitation:
20. around the Aral Sea _____
21. in the Tien Shan _____
22. along the Arctic Ocean coastline _____
23. along the axis of the Zagros Mountains _____
24. along the southwest coastline of India _____

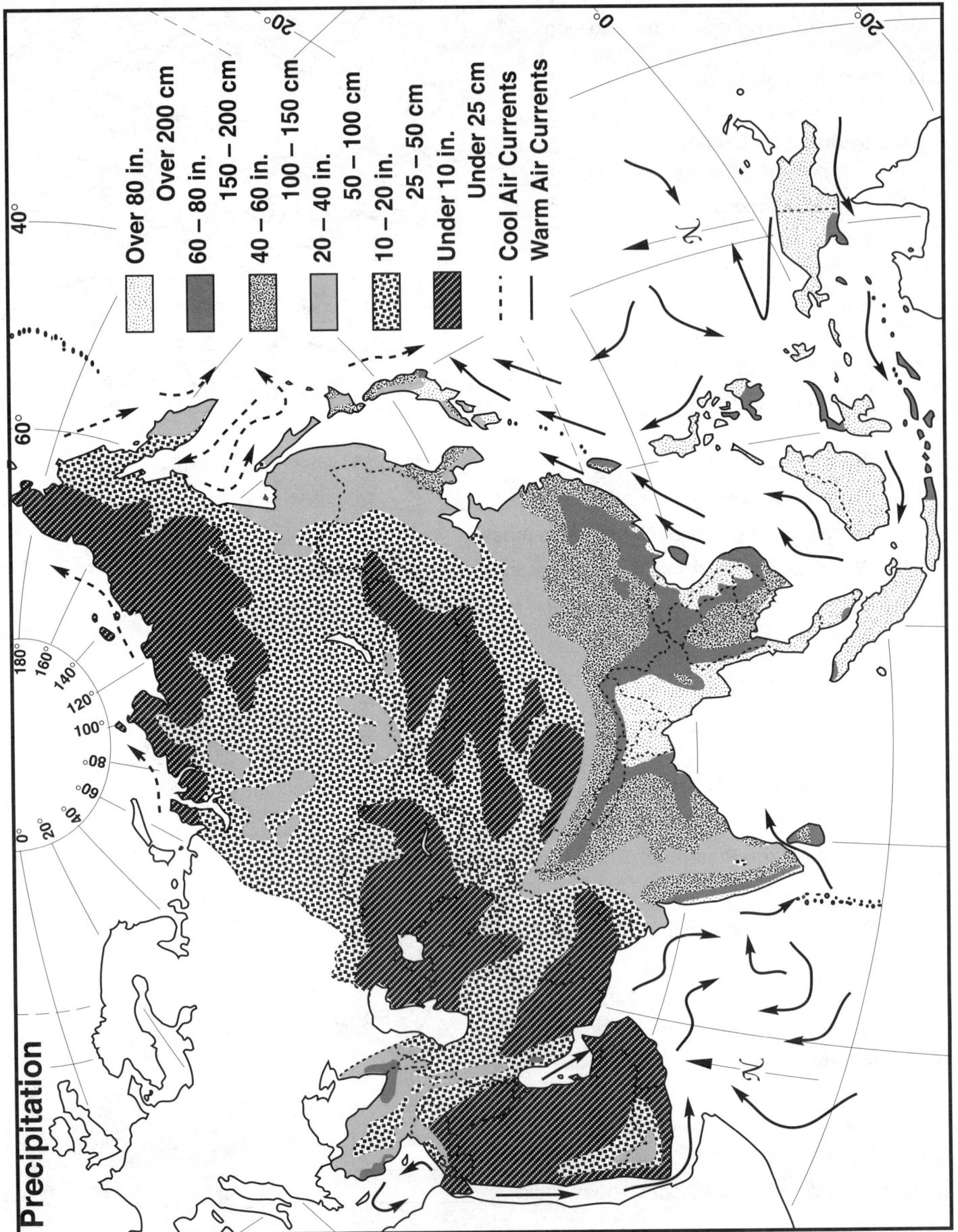

# Precipitation

| | |
|---|---|
| Over 80 in. | Over 200 cm |
| 60 – 80 in. | 150 – 200 cm |
| 40 – 60 in. | 100 – 150 cm |
| 20 – 40 in. | 50 – 100 cm |
| 10 – 20 in. | 25 – 50 cm |
| Under 10 in. | Under 25 cm |

- - - Cool Air Currents
——— Warm Air Currents

Map Skills — Asia

# Climates

Use the maps to complete the following.

1. What kind of climate is found along the southwest coast of India? _____

   How much rainfall does that area receive? _____

2. What is the climate type for the northernmost sections of Russia? _____

   How much rainfall does most of that area receive? _____

3. What is the climate type for the area around the Aral Sea? _____

   How much rainfall does that area receive? _____

4. Compare the rainfall amounts for questions 2 and 3. Why are the climate types different?

   _____

5. What are the two climate types in Japan? _____

6. Most of Mongolia is what climate type? _____

7. What is the major climate type along the Mediterranean and Black seas? _____

   At what approximate latitude is this climate type found? _____

8. Virtually all of the tropical rainy climate areas are found between the Tropic of

   _____ at 23.5°N and the Tropic of _____ at 23.5°S.

9. Name the climate type surrounding Lake Baikal. _____

10. What two climate types are found on the island of Borneo (115°E, 0°S)? _____

    _____

---

List the climate type for these cities.

11. Singapore _____

12. Beijing _____

13. Kathmandu _____

14. Hanoi _____

15. Riyadh _____

16. Islamabad _____

Match the country and the dominant climate type.

17. _____ Indonesia    a. highland
18. _____ Bhutan       b. humid continental
19. _____ North Korea  c. humid subtropical
20. _____ Pakistan     d. mediterranean
21. _____ Taiwan       e. tropical rainy
22. _____ Lebanon      f. tropical semi-arid

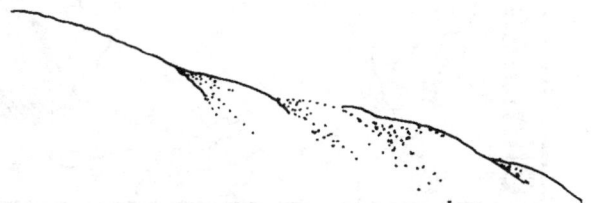

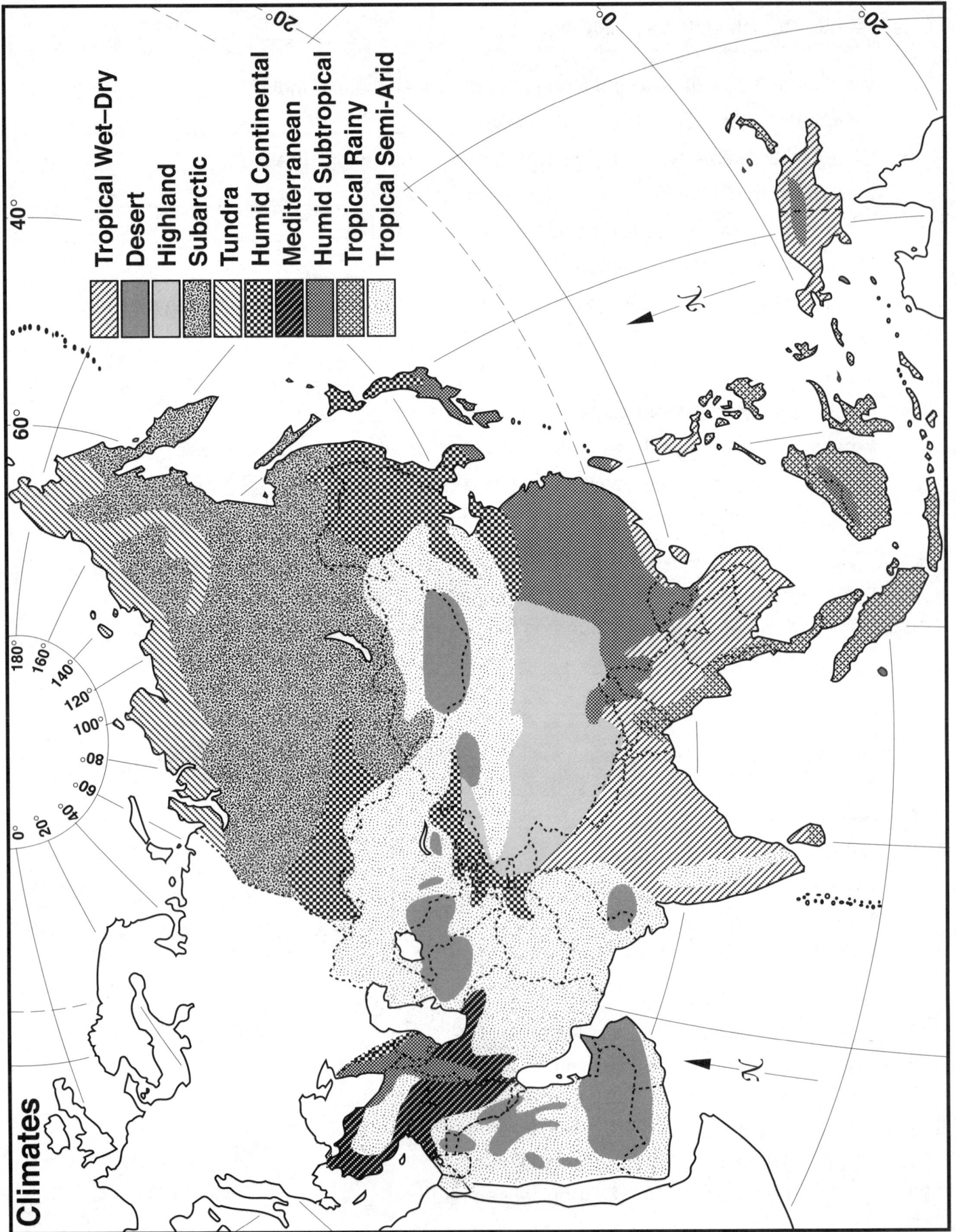

# Climates

**Legend:**
- Tropical Wet–Dry
- Desert
- Highland
- Subarctic
- Tundra
- Humid Continental
- Mediterranean
- Humid Subtropical
- Tropical Rainy
- Tropical Semi-Arid

40°  60°

20°  0°  20°

180° 160° 140° 120° 100° 80° 60° 40° 20° 0°

N

N

Use the maps to complete the following.

1. The Plateau of Tibet is either barren or has what type of vegetation? _____

2. Most of the Indus River Valley is what vegetation type? _____

3. Name three vegetation types found in India. _____

4. Rainforests require at least how much rainfall? _____

5. What climate type is associated with rainforests? _____

6. What climate types are associated with barren land? _____

7. Most of the island of New Guinea is characterized as _____

   and receives _____ cm/in. of rain annually. Why does the central part of New Guinea

   have a different type of vegetation from the coastal regions? _____

8. What climate type is associated with most mediterranean vegetation types? _____

9. The Mekong River drains an area that is virtually all _____ and

   _____ vegetation types.

10. From the Himalayas, the Ganges River flows _____ to the _____ of

    _____ through areas that are mostly _____ vegetation.

11. The subarctic climate areas of Russia are covered by extensive coniferous

    forests called _____ .

12. Most of Mesopotamia is of the _____ vegetation type. Areas to the south-

    west of Mesopotamia are mostly _____ while areas to the northwest and east

    are of the _____ type.

13. What three vegetation types are found in Turkey? _____

    _____

14. Southern Japan has some of the northernmost rainforests in Asia. Why do you think that

    rainforests can exist further north in this area than elsewhere in Asia? _____

    _____

15. The Yenisey River flows _____ from the Russia/Mongolia border through regions

    of _____ and _____ vegetation.

Match the city and the dominant vegetation type.

16. _____ Kuala Lumpur    a. mediterranean
17. _____ Riyadh          b. deciduous
18. _____ Kabul           c. desert
19. _____ New Delhi       d. rainforest
20. _____ Beirut          e. dry savannah

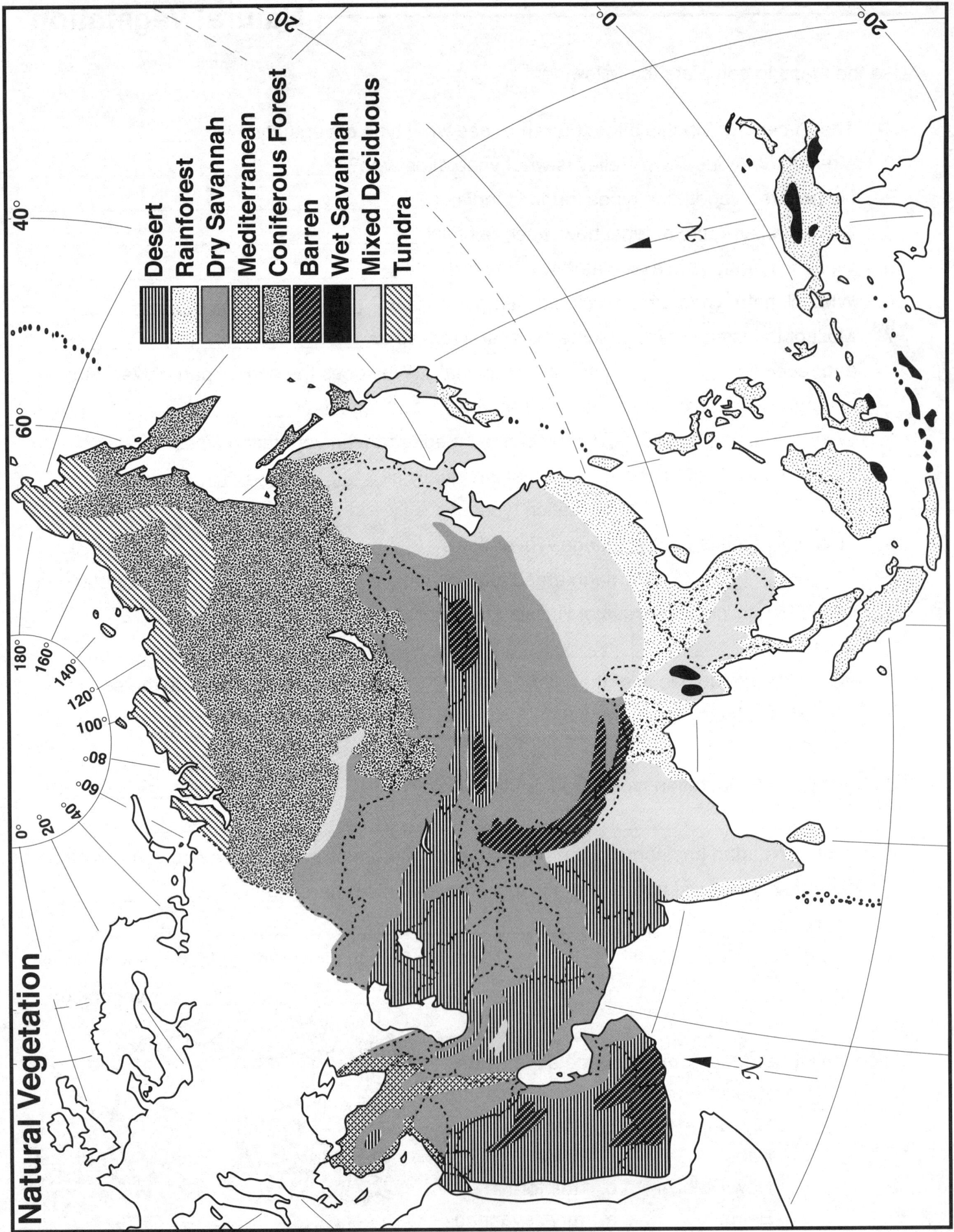

# Natural Vegetation

**Legend:**
- Desert
- Rainforest
- Dry Savannah
- Mediterranean
- Coniferous Forest
- Barren
- Wet Savannah
- Mixed Deciduous
- Tundra

40°
20°
0°
20°

60°

180°
160°
140°
120°
100°
80°
60°
40°
20°
0°

N

N

# Products

Use the maps and additional references to complete the crossword puzzle.

## Across

3. Wheat, rice, and jute are produced in ____ . It borders Afghanistan and India.
8. ____ is located at 20°N, 80°E. It produces cotton, wheat, and rice.
9. The Hindu Kush is in this mountainous country. Little land is under cultivation.
10. ____ is on the Red Sea. It produces dates and coffee, and its capital is Sanaa.
11. ____ is one of the world's leading producers of chromium. Its capital is Ankara.
13. Copper and chromium are minerals found in the Philippines. Its capital is ____ .

## Down

1. ____ is an island located east of China. Rice is grown there.
2. The capital of ____ is Tokyo. It produces zinc, coal, manganese, rice, and tea.
4. The former ____ is a leading producer of wheat. It also produces tin, zinc, and lead.
5. Myanmar is the northernmost country of Southeast Asia. It grows rice and ____ .
6. Saudi Arabia is the world's leading exporter of ____ . Its capital is Riyadh.
7. ____ has the largest population in the world. It is a major producer of tungsten.
8. The capital of ____ is Jakarta. It produces rubber, corn, tin, nickel, and petroleum.
12. Seoul is the capital of South Korea. ____ is one of its main crops.
14. Petroleum is a major natural resource of ____ , located east of Iraq.

# Products

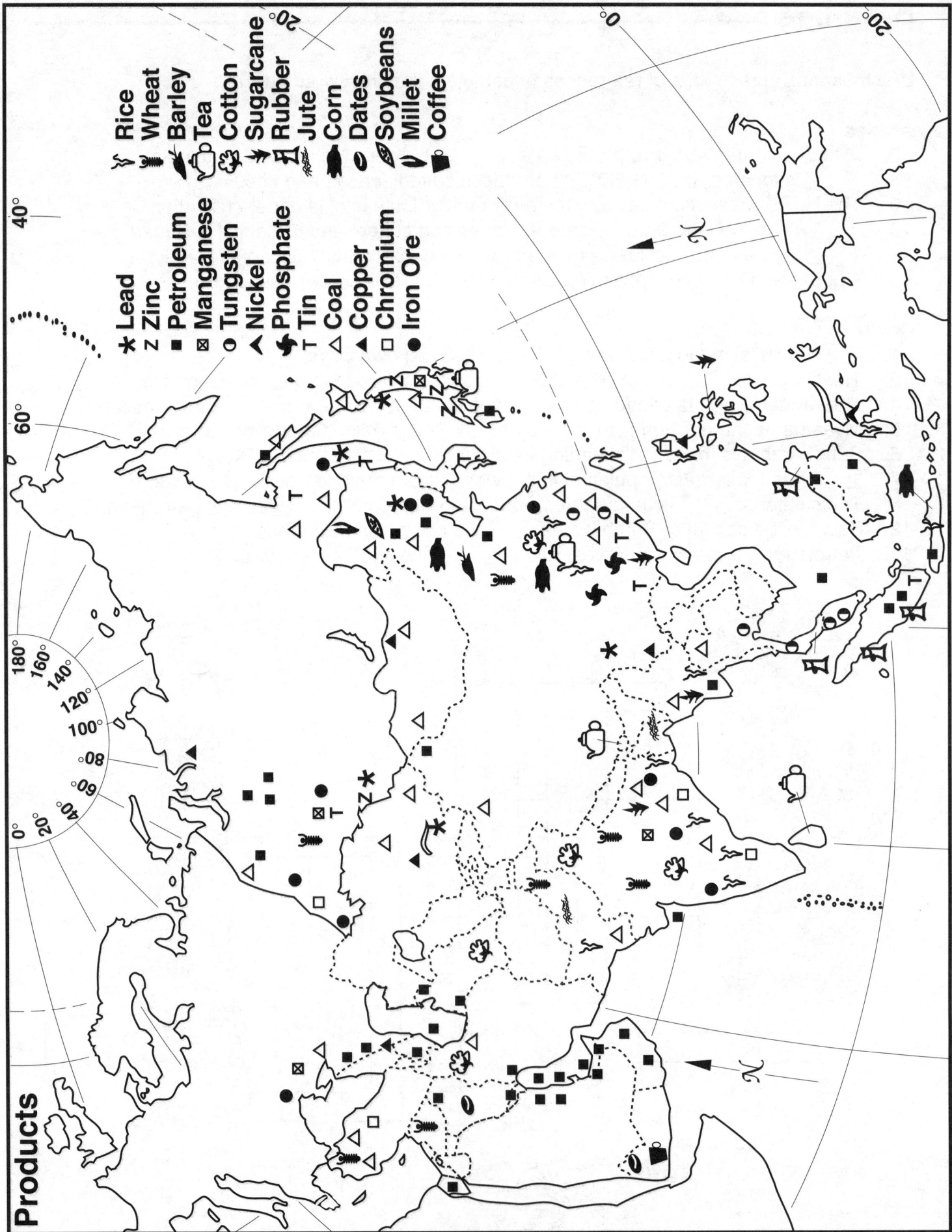

**Legend (crops):**
- Rice
- Wheat
- Barley
- Tea
- Cotton
- Sugarcane
- Rubber
- Jute
- Corn
- Dates
- Soybeans
- Millet
- Coffee

**Legend (minerals):**
- ✳ Lead
- z Zinc
- ■ Petroleum
- ⊠ Manganese
- ○ Tungsten
- ◀ Nickel
- ⌐ Phosphate
- T Tin
- △ Coal
- ◆ Copper
- □ Chromium
- ● Iron Ore

Use the maps and additional references to answer the following questions.

1. What is most of the land in the Gobi Desert used for? _____
   What is its climate type? _____

2. Just to the south of the Ural Mountains is the Kirghiz _____ .
   What is it used for? _____

3. What rivers flow through the farming lands of eastern China? _____
   Name three agricultural products of this region. _____

4. What problems might impede farming in Indonesia? _____

5. What two land uses are common in the Philippines? _____
   Name two products of this region. _____

6. What are the common land uses in central Turkey? _____

7. What is the land used for in south–central Saudi Arabia? _____
   Why? _____

8. Give two reasons why there is little agriculture in the areas just north of India. _____
   _____

9. What types of animals are raised by nomadic tribes? _____

10. Name two products of the rainforest regions of Indonesia. _____

11. Name the land use areas of southeast Asia where most of the petroleum is found. _____
    _____

12. Some farming is done in areas that are used primarily for herding. What limits the
    people's ability to use more of this land for farming? _____
    _____

13. Why is it important that most of India and eastern China be used for farming? _____
    _____

14. What is the difference between grazing and nomadic herding? _____
    _____

15. Which land use produces the most food—farming or grazing? _____

# Land Use

**Legend:**
- Farming
- Forest (Hunting and Gathering)
- Nomadic Herding
- Little Used Land
- Stock Grazing
- Fishing

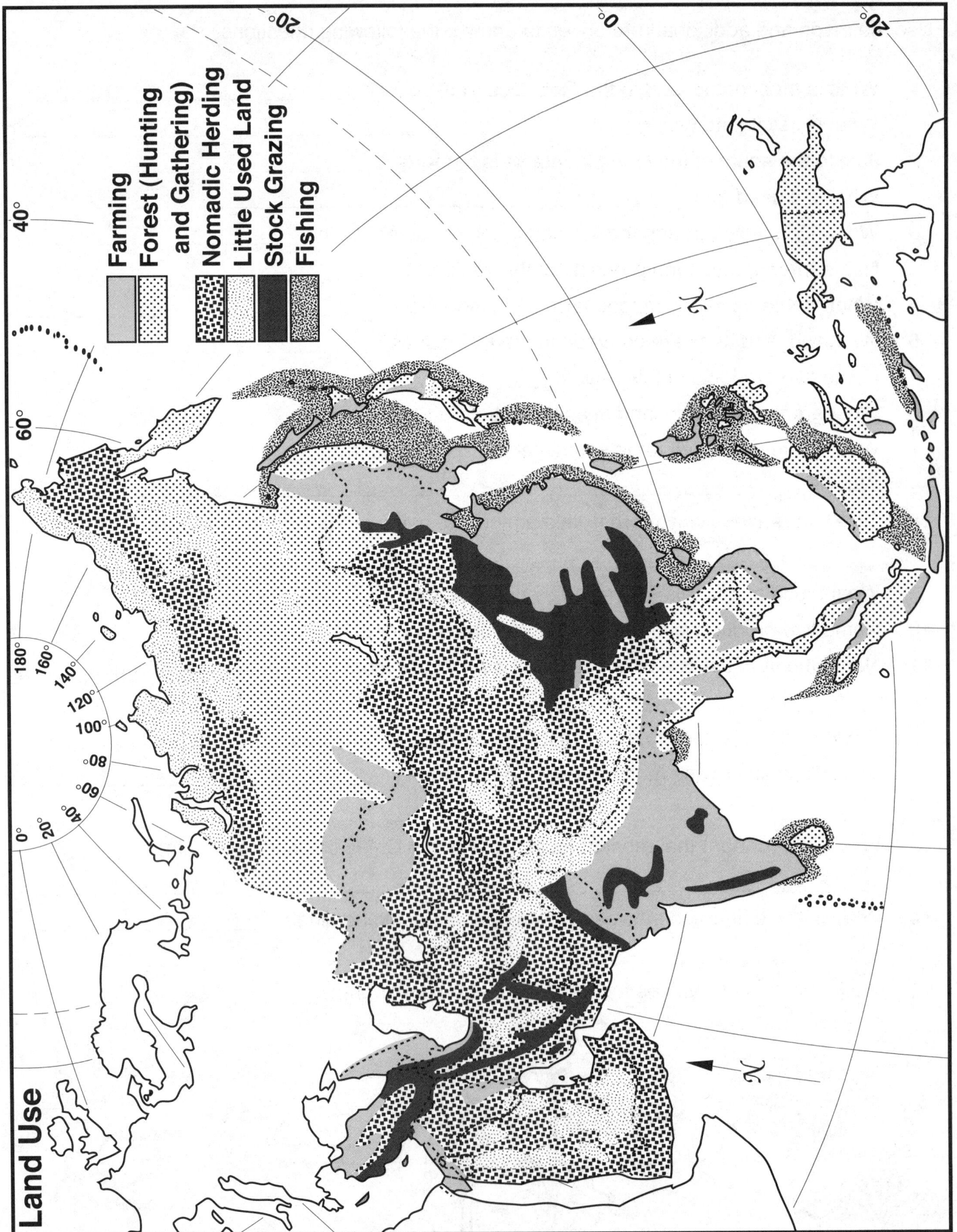

# Population Patterns and Large Cities ───────────

Use the maps to complete the following.

Label these Asian cities: Ankara, Tehran, Chennai (Madras), Mumbai (Bombay), Delhi, Kolkata (Calcutta), Karachi, Tokyo, Osaka, Taipei, Seoul, Busan, Singapore, Ho Chi Minh City, Bangkok, Yangon (Rangoon), Manila, Tashkent, Jakarta, Surabaya, Shenyang, Harbin, Tianjin, Beijing, Shanghai, Nanjing, Wuhan, Hong Kong, Chongqing, Chengdu, Xi'an, Taiyuan, Islamabad, Jerusalem, Ulaanbaatar, Kunming, Hanoi, Nanchang, Qingdao, Dalian, Sapporo, Sendai, Kyoto, and Hiroshima.

1. Name two countries which have large areas of dense population. _____

2. What is the population density for most of the Plateau of Tibet?_____
   Why is this? _____

3. What factors keep the population density low in much of Saudi Arabia? _____
   _____

4. The Ob, Yenisey, and Lena river valleys have population densities higher than the
   surrounding areas. Why?_____

5. Where do most of the people in the Philippines live? _____

**Across**
1. the "new" city in India
4. city-state on the Malay Peninsula
6. now called Ho Chi Minh City
7. capital of Turkey
9. Pakistani port on the Arabian Sea
10. The 1988 Summer Olympics were held in this Korean city.

**Down**
2. long-time British colony in China (two words)
3. capital city east of the South China Sea
4. large port city on the East China Sea
5. capital of Japan
8. This was the old name for the city of Chennai.

# Population Patterns and Large Cities

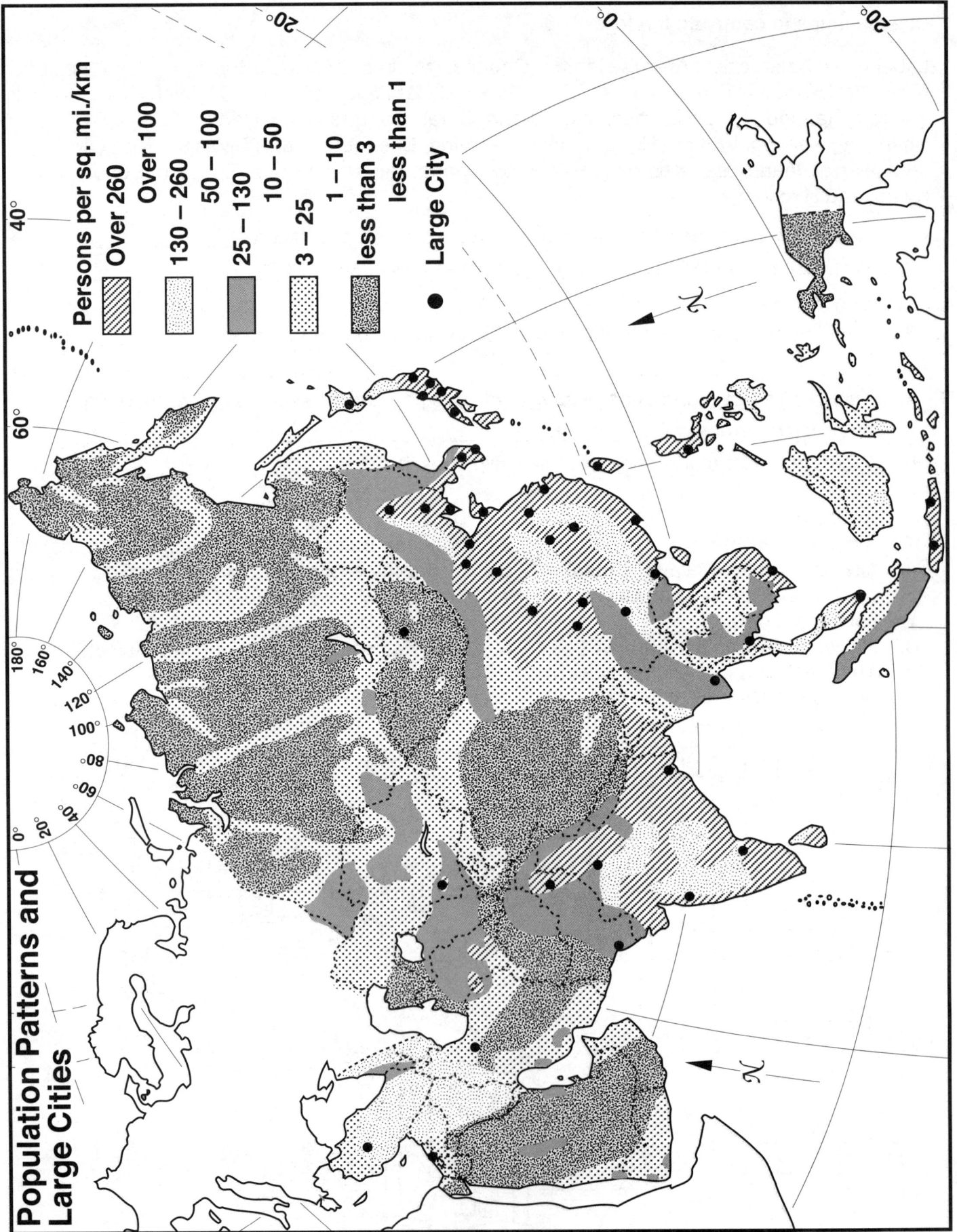

**Persons per sq. mi./km**

| | |
|---|---|
| Over 260 | Over 100 |
| 130 – 260 | 50 – 100 |
| 25 – 130 | 10 – 50 |
| 3 – 25 | 1 – 10 |
| less than 3 | less than 1 |
| ● | Large City |

Label the following **boldfaced** civilizations on the map.

**Before 1000 B.C.**  The earliest civilizations in Asia were also some of the earliest that existed in the world. Shown are the approximate boundaries of ancient **Mesopotamia** (Greek for "land between the waters"), including Sumeria and Babylonia. This land is located in what is now Iraq and Iran.  A civilization based on early city–states was found along the **Indus River Valley** in modern Pakistan.  Another civilization was based along the **Huang He River.**

**1000 B.C.– 0**  Many empires flourished during this era.  In southwest Asia, the mighty **Persian Empire** stretched from Egypt in Africa to southeastern Europe and east to the Indus River Valley.  The map represents the greatest extent of the empire under King Darius. In what is now India, the **Mauryan Civilization** is shown under its great King Asoka. The ancient Chinese **Ch'in Dynasty** boundaries are shown in what is now modern China.

**0–500 A.D.**  In western Asia, two mighty empires existed at approximately the same time: the **Roman Empire** and the **Persian Empire.**  The boundary between these (roughly 100 A.D. under the Roman emperor Trajan) ran from the Caucasus Mountains through Mesopotamia to the Syrian Desert. Bordering the Persians to the east was the mountain empire of the **Kushans.** India was home to the **Gupta Civilization,** and China is represented by the **Han Dynasty.**

**500–1000 A.D.**  In western Asia, Africa, and Europe, the **Muslim Empire** rose to prominence and included much of the old Roman and Persian Empires. A distinct civilization in southeast Asia, the **Khmer,** is shown, as is the **Heian Empire** of Japan.

**1000–1600 A.D.**  The map represents two important civilizations of this time period: the **Moguls,** who ruled India for several centuries, and the **Ming Dynasty** of China.

1. Name three dynasties of ancient China. _____

2. Name three modern Asian countries that were included in the Persian Empire. _____

    _____

3. Name three empires that controlled the land occupied by the current country of Israel.

    _____

4. Why do you suppose that Japan was never occupied by an outside empire such as China? _____

Match the ancient civilization and the modern country.
| | | | |
|---|---|---|---|
| 5. | _____ | Gupta | a. Saudi Arabia |
| 6. | _____ | Muslim | b. Afghanistan |
| 7. | _____ | Roman | c. Thailand |
| 8. | _____ | Kushan | d. Turkey |
| 9. | _____ | Khmer | e. North Korea |
| 10. | _____ | Han | f. India |

# Civilzations B.C.

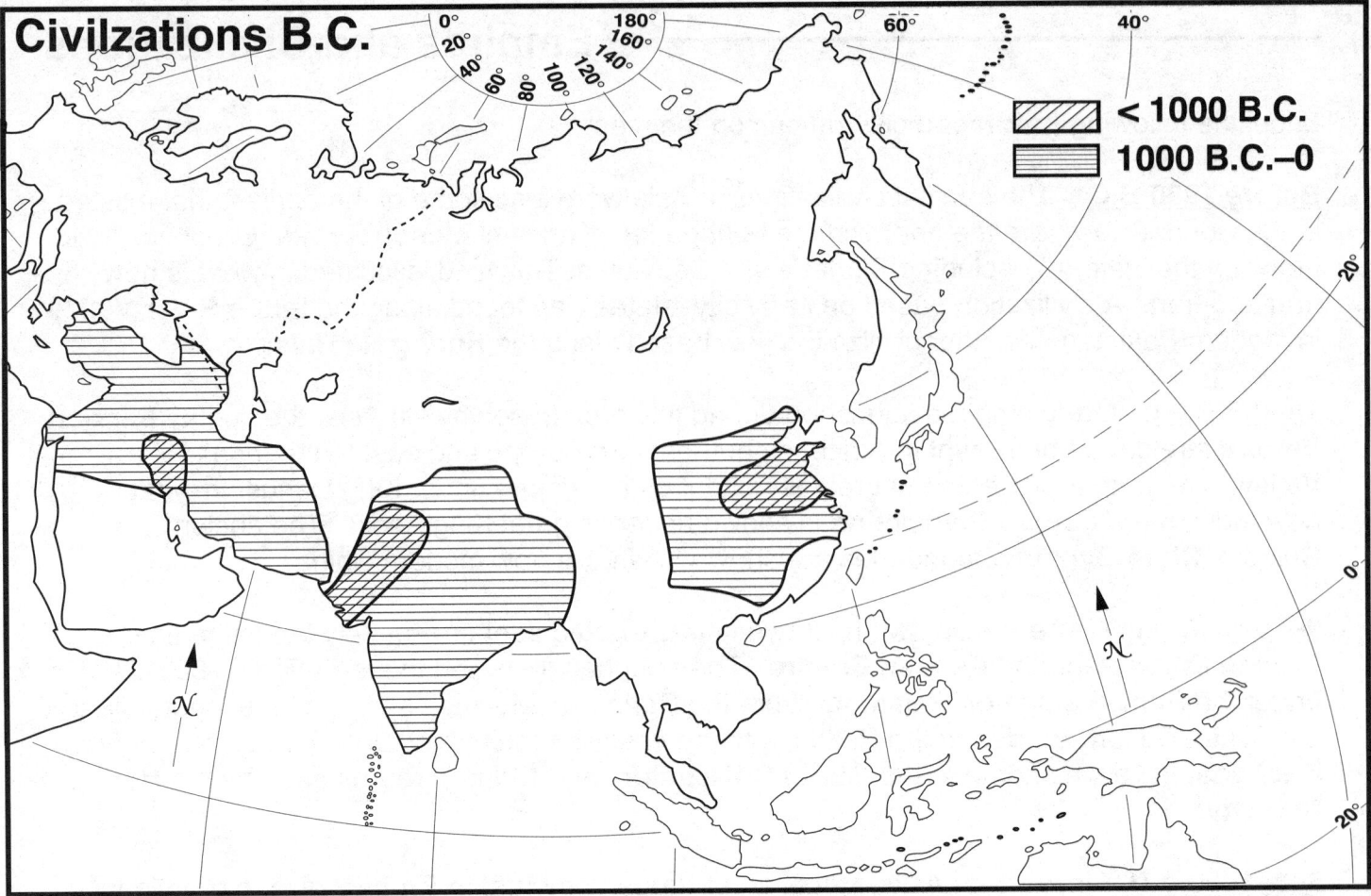

| | |
|---|---|
| ▨ | < 1000 B.C. |
| ▤ | 1000 B.C.–0 |

0° 20° 40° 60° 80° 100° 120° 140° 160° 180° 60° 40°

20° 0° 20°

N

N

# Civilzations A.D.

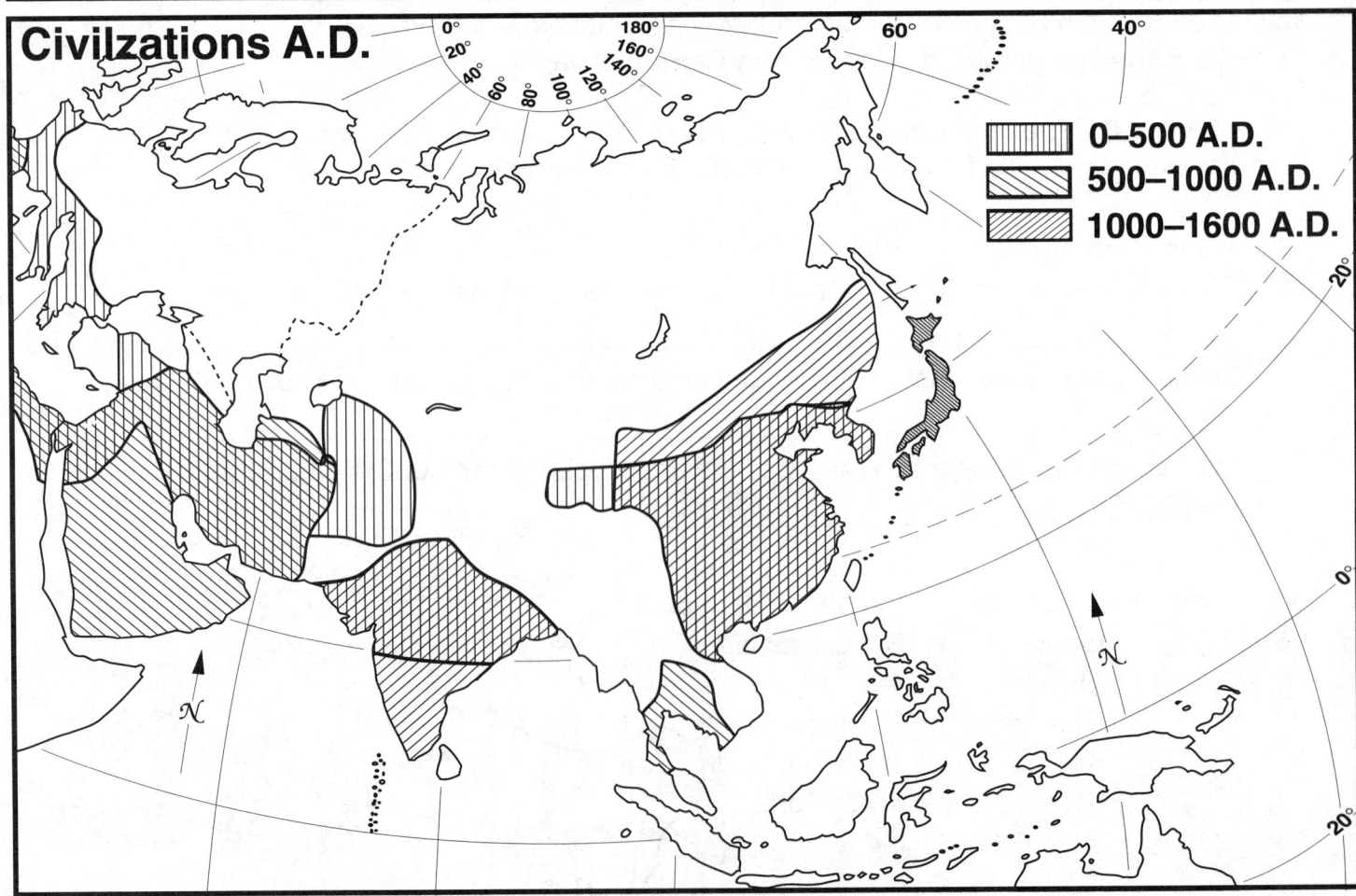

| | |
|---|---|
| ▥ | 0–500 A.D. |
| ▨ | 500–1000 A.D. |
| ▨ | 1000–1600 A.D. |

0° 20° 40° 60° 80° 100° 120° 140° 160° 180° 60° 40°

20° 0° 20°

N

N

11a     Map Skills — Asia

# The Former Soviet Union

Use previous maps or additional references to label the features and to complete the following.

1. Label the countries that made up the former Soviet Union and their capitals.
2. Label the cities.
3. Label these features:  Kara–Kum Desert, Kamchatka Peninsula, Ural Mountains, Caucasus Mountains, Kirghiz Steppe, Sakhalin, Tien Shan, Central Siberian Plateau, and West Siberian Plain.
4. Label these bodies of water:  Ural River, Black Sea, Volga River, Sea of Okhotsk, Arctic Ocean, Lake Balkhash, Amu Darya, Baltic Sea, Ob River, Lena River, Yenisey River, Lake Baikal, Aral Sea, Caspian Sea, Amur River, and Sea of Japan.

5. Use latitudes and longitudes to label these cities on the map.

| City | Latitude | Longitude |
|------|----------|-----------|
| St. Petersburg | 59.57°N | 30.20°E |
| Vladivostok | 43.06°N | 131.47°E |
| Samarkand | 39.50°N | 68.00°E |
| Irkutsk | 52.16°N | 104.00°E |
| Omsk | 55.12°N | 73.19°E |
| Novosibirsk | 55.09°N | 82.58°E |
| Ulan Ude | 52.00°N | 111.50°E |

6. Compare these cities by completing the chart.

| City | Climates | Land Uses | Products |
|------|----------|-----------|----------|
| Omsk | | | |
| Vladivostok | | | |
| Tashkent | | | |
| Baku | | | |

Complete the following statements with directions, distances, and features.

7. To get to the Pacific Ocean from Vladivostok, a ship must cross the Sea of _____ and either the _____ Sea or the Sea of _____ .

8. Tashkent lies _____ km/mi. _____ of Novosibirsk.

9. The Lena River mostly flows _____ into the _____ .

10. The Ural Mountains run in a _____ to _____ direction.

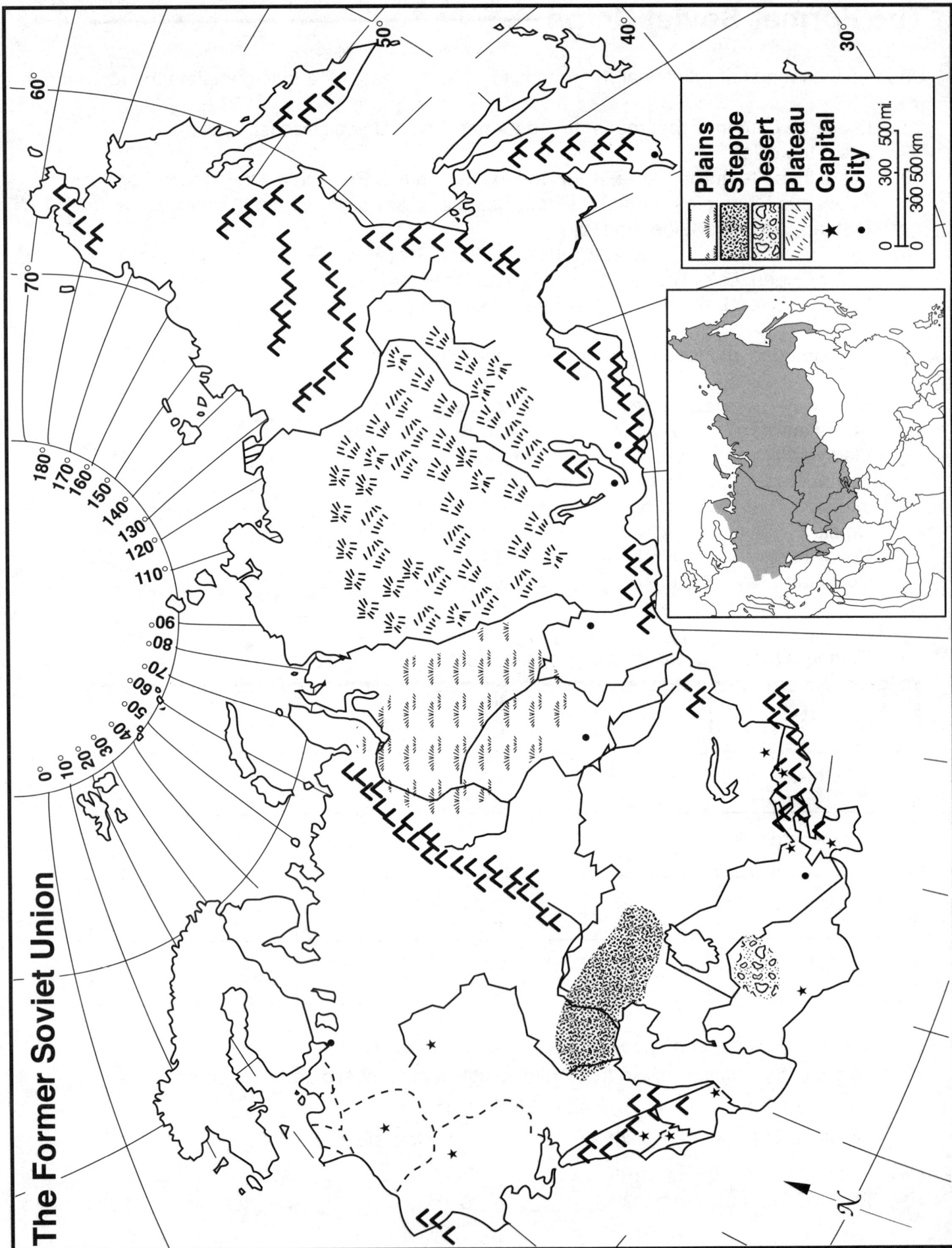

# The Former Soviet Union

Plains
Steppe
Desert
Plateau
Capital
City

300  500 mi.

300  500 km

60°
70°
50°
40°
30°

180°
170°
160°
150°
140°
130°
120°
110°
90°
80°
70°
60°
50°
40°
30°
20°
10°
0°

N

# Southwestern Asia

Use previous maps or additional references to label the features and to complete the following.

1. Label the countries and capitals on the map.
2. Label these features: Iranian Plateau, Zagros Mountains, Caucasus Mountains, Rub al Khali, Anatolian Plateau, Syrian Desert, and Sinai Peninsula.
3. Label these bodies of water: Gulf of Aden, Strait of Hormuz, Sea of Marmara, Aegean Sea, Dardanelles, Black Sea, Gulf of Oman, Red Sea, Suez Canal, Dead Sea, Mediterranean Sea, Arabian Sea, Caspian Sea, Persian Gulf, Euphrates River, and Tigris River.

4. Use latitudes and longitudes to label the cities on the map.

| City | Country | Latitude | Longitude |
|------|---------|----------|-----------|
| Istanbul | _____ | 41.02°N | 29.00°E |
| Tel Aviv | _____ | 32.01°N | 34.30°E |
| Haifa | _____ | 32.48°N | 35.00°E |
| Tabriz | _____ | 38.00°N | 46.13°E |
| Esfahan | _____ | 32.38°N | 51.30°E |
| Medina | _____ | 24.26°N | 39.42°E |
| Mecca | _____ | 21.27°N | 39.45°E |
| Aleppo | _____ | 36.10°N | 37.18°E |

5. Compare these countries by completing the chart.

| Country | Climates | Land Uses | Products |
|---------|----------|-----------|----------|
| Yemen | | | |
| Iran | | | |
| Saudi Arabia | | | |
| Israel | | | |

Complete the following statements with directions, distances, and features.

6. To travel from Istanbul to Kuwait by boat, one would sail across the Sea of _____ through the straits called the _____ , across the _____ and _____ seas, through the _____ Canal, down the length of the _____ Sea, through the Gulf of _____ , across portions of the _____ Sea, through the Gulf of _____ and the Strait of _____ , and then across the _____ Gulf.

7. The capital of Jordan lies _____ km/mi. _____ of Damascus.

8. When traveling from Yemen to Bahrain, one would have to cross the _____ .

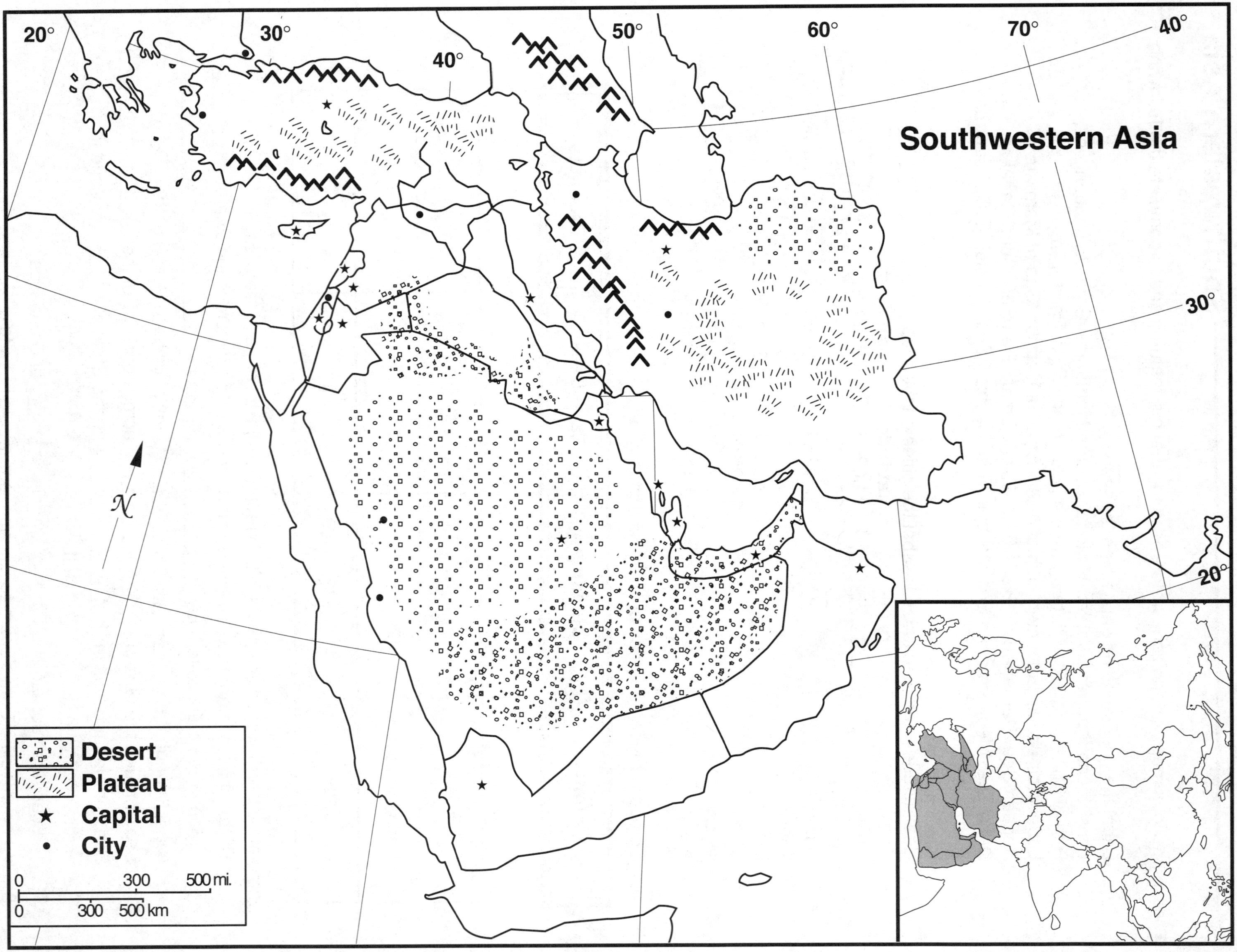

# Southwestern Asia

40°

30°

20°

70°

60°

50°

40°

30°

20°

N

13a

Map Skills — Asia

Desert

Plateau

Capital ★

City •

300   500 mi.

300   500 km

300   500 km

# Southeastern Asia

Use previous maps or additional references to label the features and to complete the following.

1.  Label the countries and capitals on the map.
2.  Label these features:  Sumatra, Borneo, Luzon, New Guinea, Malay Peninsula, Maoke Mountains, Java, and Mindanao.
3.  Label these bodies of water:  Strait of Malacca, Gulf of Thailand, Celebes Sea, Gulf of Tonkin, Banda Sea, Sulu Sea, Arafura Sea, Andaman Sea, Irrawaddy River, Salween River, Mekong River, Philippine Sea, Java Sea, and South China Sea.

4.  Use latitudes and longitudes to label these cities on the map.

| City | Country | Latitude | Longitude |
|---|---|---|---|
| Medan | _____ | 3.35°N | 98.35°E |
| Surabaya | _____ | 7.23°S | 112.45°E |
| Bandung | _____ | 7.00°S | 107.22°E |
| Manado | _____ | 1.29°N | 124.50°E |
| Mandalay | _____ | 22.00°N | 96.08°E |
| Da Nang | _____ | 16.08°N | 108.22°E |
| Ho Chi Minh City | _____ | 10.46°N | 106.34°E |

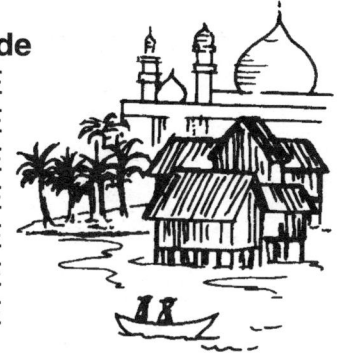

5.  Compare these countries by completing the chart.

| Country | Climates | Land Uses | Products |
|---|---|---|---|
| Indonesia | | | |
| Malaysia | | | |
| Cambodia | | | |
| Philippines | | | |
| Myanmar | | | |

Complete the following statements with directions, distances, and features.

6.  To go from Nay Pyi Taw to Surabaya by sea, a ship would sail _____ through the _____ Sea, through the Strait of _____ , past the city–state of _____ , and across the _____ Sea.

7.  Hanoi lies _____ km/mi. _____ of Ho Chi Minh City.

8.  To fly from Bangkok, the capital of _____ , to _____, the capital of the Philippines, one would fly _____ for _____ km/mi. and cross the _____ _____ Sea.

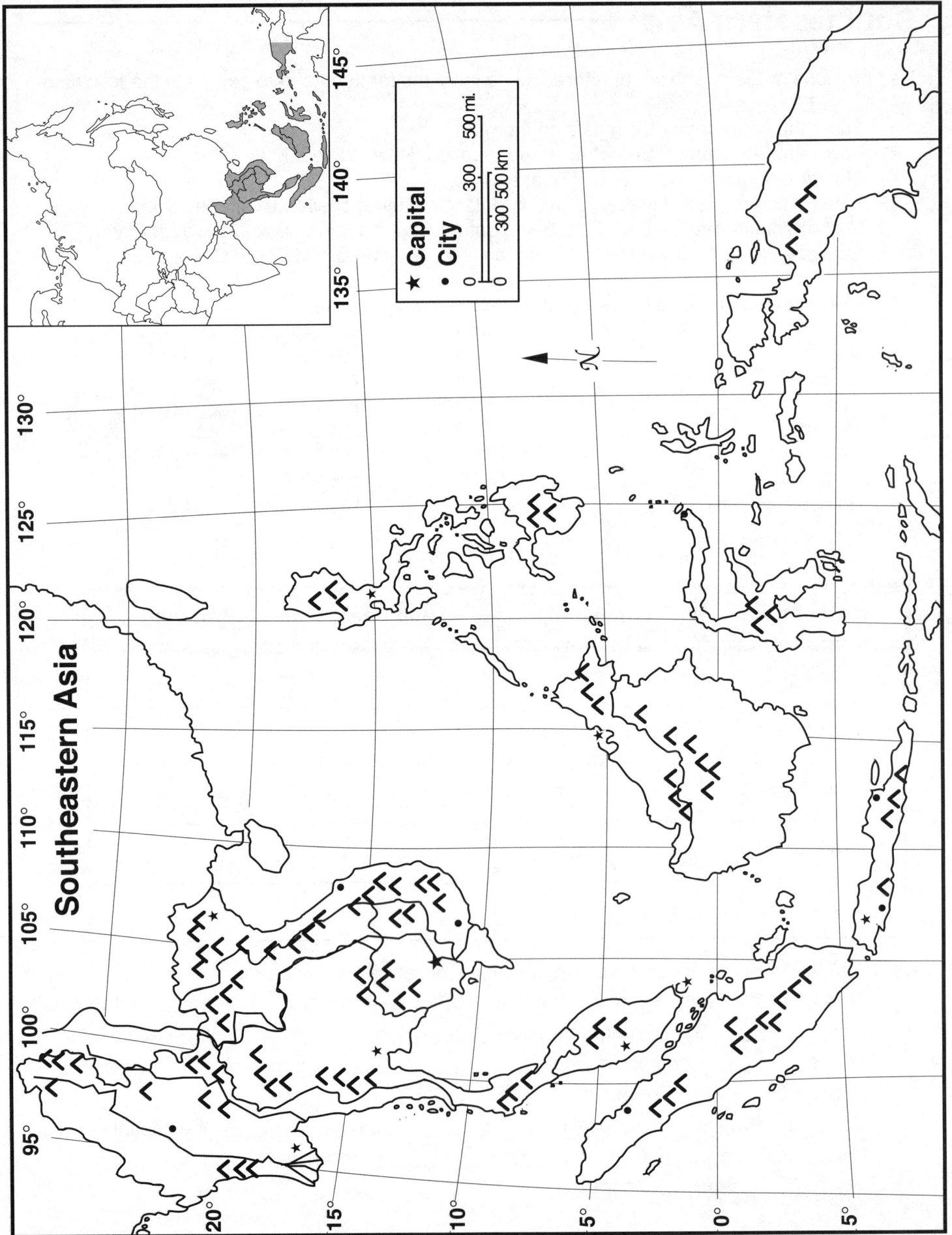

# Southeastern Asia

145°
140°
135°
130°
125°
120°
115°
110°
105°
100°
95°

★ **Capital**
● **City**

500 mi.
300
300 500 km
0
0

N

20°
15°
10°
5°
0°
5°

14a

# Eastern Asia

Use previous maps or additional references to label the features and to complete the following.

1. Label the countries, capitals, and cities on the map.
2. Label these features: the Himalayas, Tien Shan, Honshu, Plateau of Tibet, Gobi Desert, Shandong Peninsula, Hokkaido, Kunlun Mountains, and Okinawa.
3. Label these bodies of water: Taiwan Strait, Sea of Japan, East China Sea, South China Sea, Yellow Sea, Amur River, Huang He River (Yellow River), Xi River, Yangtze River, and Mekong River.

4. Use latitudes and longitudes to label these cities on the map.

| City | Country | Latitude | Longitude |
|---|---|---|---|
| Busan | _____ | 35.08°N | 129.05°E |
| Shanghai | _____ | 31.14°N | 121.27°E |
| Hong Kong | _____ | 21.45°N | 115.00°E |
| Lhasa | _____ | 29.41°N | 91.12°E |
| Xi'an | _____ | 34.20°N | 109.00°E |
| Zhengzhou | _____ | 34.46°N | 113.42°E |
| Yokohama | _____ | 35.37°N | 139.40°E |
| Osaka | _____ | 34.45°N | 135.36°E |

5. Compare these countries by completing the chart.

| Country | Climates | Land Uses | Products |
|---|---|---|---|
| China | | | |
| Mongolia | | | |
| South Korea | | | |
| Japan | | | |

Complete the following statements with directions, distances, and features.

6. Flying from Beijing to Ulaanbaatar would take one in a _____ direction for _____ km/mi. and over the _____ Desert.
7. To go from Dalian to Taipei, a ship would sail in a _____ direction and pass through the _____ Sea and the _____ Sea.
8. Lhasa lies in the _____ of _____ . To fly from Lhasa to Shanghai, one would go mostly _____ for _____ km/mi.

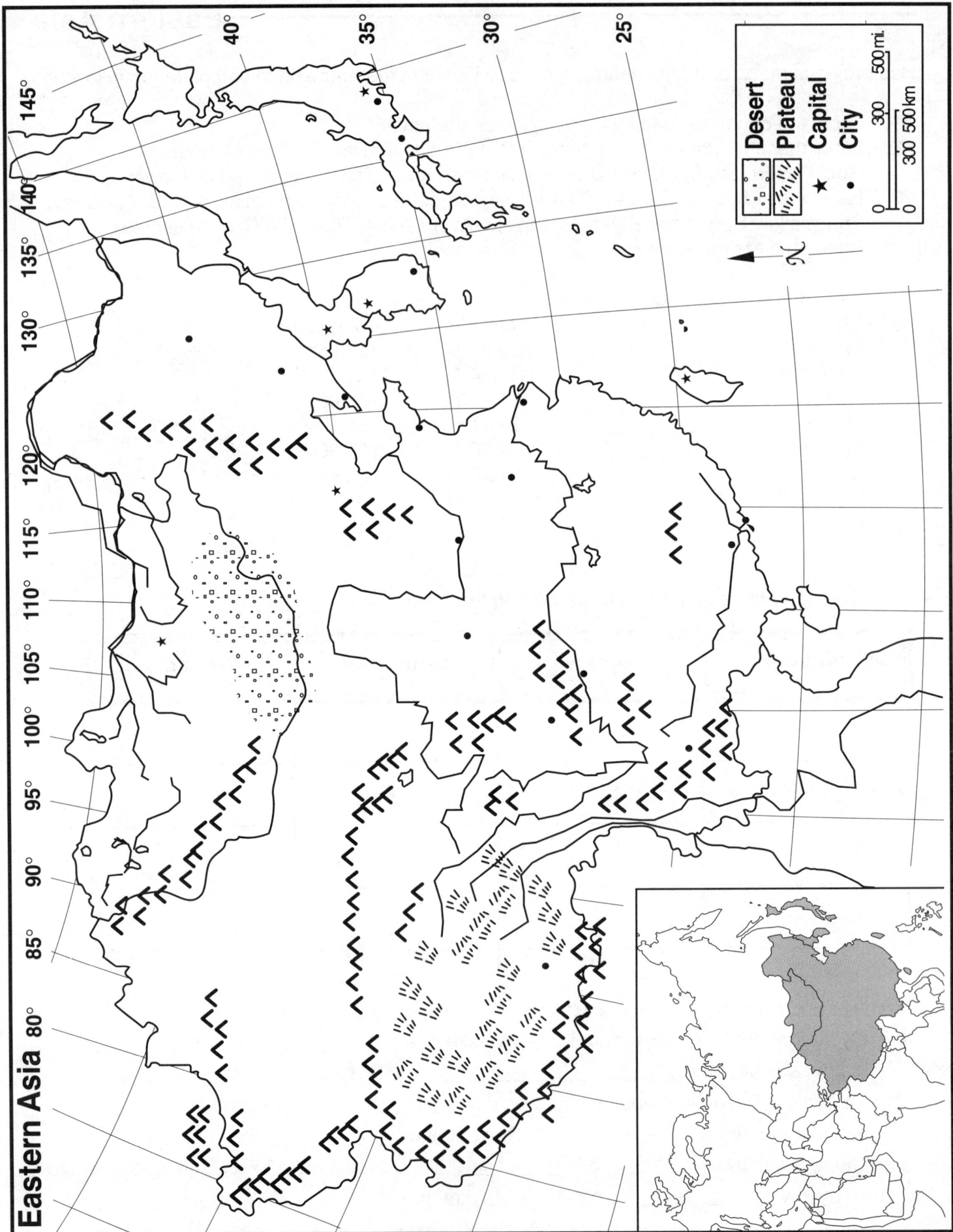

# Eastern Asia

145°
140°
135°
130°
120°
115°
110°
105°
100°
95°
90°
85°
80°

40°
35°
30°
25°

**Desert**
**Plateau**
**Capital** ★
**City** ●

500 mi.
300
300 500 km
0

N

Map Skills — Asia

# Southern Asia

Use previous maps or additional references to label the features and to complete the following.

1. Label the countries and capitals on the map.
2. Label these features: Hindu Kush, the Himalayas, Great Indian Desert, Deccan Plateau, Karakoram Range, Western Ghats, and Khyber Pass.
3. Label these bodies of water: Brahmaputra River, Ganges River, Indus River, Arabian Sea, Indian Ocean, and Bay of Bengal.

4. Use latitudes and longitudes to label these cities on the map.

| City | Country | Latitude | Longitude |
|------|---------|----------|-----------|
| Lahore | _____ | 32.00°N | 74.18°E |
| Karachi | _____ | 24.59°N | 68.56°E |
| Chittagong | _____ | 22.26°N | 90.51°E |
| Lalitpur | _____ | 27.23°N | 85.24°E |
| Kandy | _____ | 7.18°N | 80.42°E |
| Chennai | _____ | 13.08°N | 80.15°E |
| Kolkata | _____ | 22.32°N | 88.22°E |
| Delhi | _____ | 28.54°N | 77.13°E |
| Bangalore | _____ | 13.03°N | 77.39°E |
| Mumbai | _____ | 18.58°N | 72.50°E |

5. Compare these countries by completing the chart.

| Country | Climates | Land Uses | Products |
|---------|----------|-----------|----------|
| Nepal | | | |
| India | | | |
| Sri Lanka | | | |

Complete the following statements with directions, distances, and features.

6. Karachi lies _____ km/mi. _____ of Lahore.

7. The capital of Nepal, _____ , lies about _____ km/mi. _____ of the capital of _____ , which is New Delhi.

8. To sail from Karachi to Madras, one would go through the _____ Sea, along the northern part of the _____ Ocean, and into the Bay of _____ .

9. Flying from Kabul to New Delhi would take one _____ for _____ km/mi. and over the country of _____ .

10. The Ganges River flows through _____ and _____ and empties into the Bay of _____ .

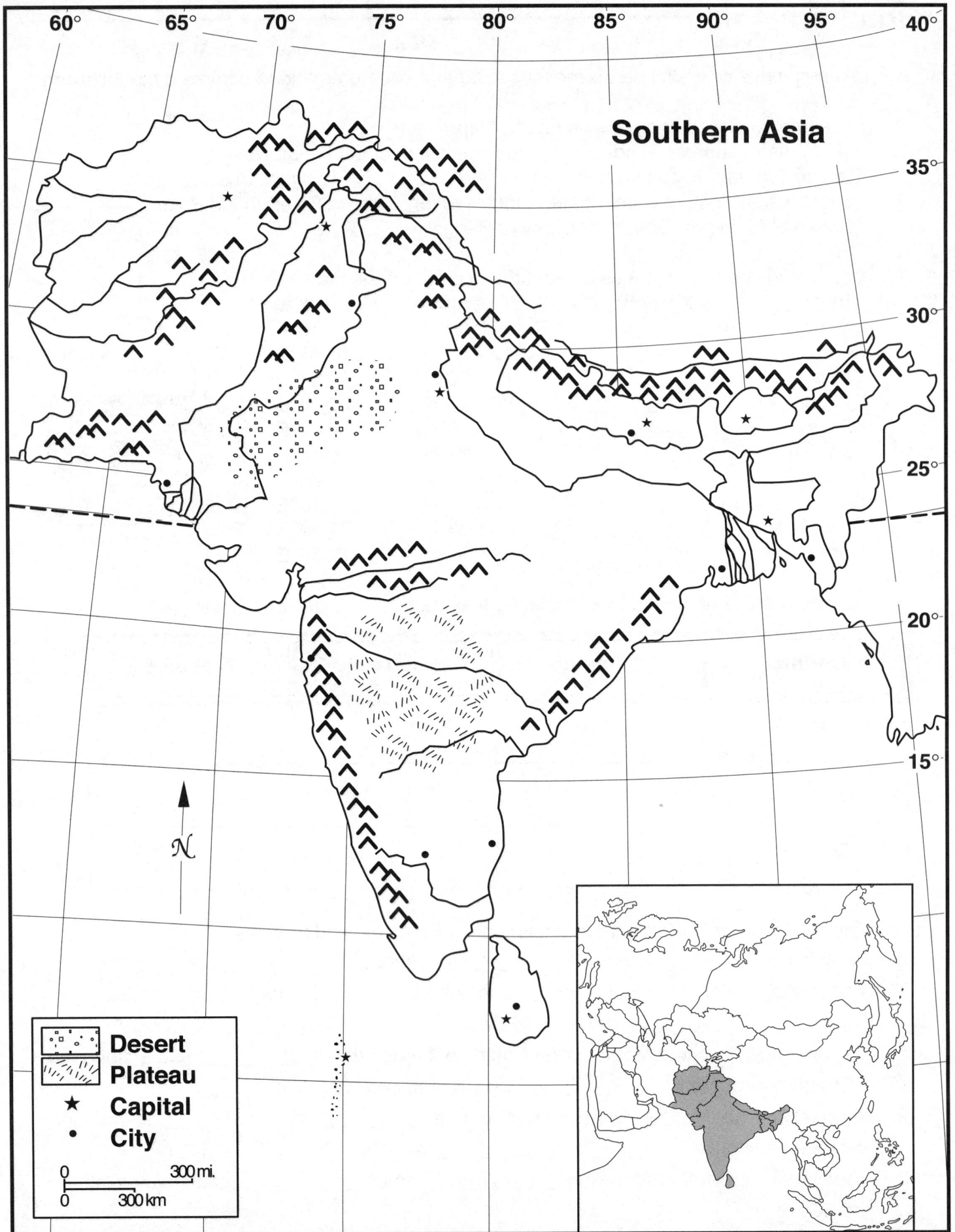

## Southern Asia

60° 65° 70° 75° 80° 85° 90° 95° 40°

35°

30°

25°

20°

15°

0         300 mi.

0         300 km

N